Illustrated
Mathematics

Illustrated Mathematics

Visualization of Mathematical Objects with Mathematica

Oliver Gloor
Beatrice Amrhein
Roman E. Maeder

Oliver Gloor
Universität Tübingen
WSI Sand 13
D-72076 Tübingen

Beatrice Amrhein
Universität Tübingen
WSI Sand 13
D-72076 Tübingen

Roman Maeder
ETH Zürich
ETH Zentrum IFW
CH-8092 Zürich

Published by TELOS, The Electronic Library of Science, Santa Clara, CA.
Publisher: Allan M. Wylde
Publishing Associate: Kate McNally Young
TELOS Production Manager: Jan Benes
Production Editor: Paul Wellin
Electronic Production: Joe Kaiping
Cover Design: Ark Stein, The Visual Group

© 1995 TELOS/Springer-Verlag
Published by TELOS, The Electronic Library Of Science, Santa Clara, California.
TELOS is an imprint of Springer-Verlag New York, Inc.

This work consists of a CD-ROM disc packaged with a booklet, and is protected by federal copyright law and international treaty. This work may not be translated or copied in whole or in part without the written permission of the publisher (Springer-Verlag New York, Inc., 175 Fifth Avenue, New York, NY 10010, USA) except for brief exerpts in connection with reviews or scholarly analysis. Use in connection with any form of information storage and retrieval, electronic adaptation computer software or by similar or dissimilar methodology now known or hereafter developed other than those expressly granted in the disc copyright and disclaimer information is forbidden.

Springer-Verlag or the authors makes no warranty of representation, either expressed or implied, with respect to this work, including its quality, merchantability, or fitness for a particular purpose. In no event will Springer-Verlag or the authors be liable for direct, indirect, special, incidental, or consequential damages arising out of the use or inability to use the disc or booklet, even if Springer-Verlag or the authors has been advised of the possibility of such damages.

MathReader contained on this CD-ROM is proprietary software created and owned by Wolfram Research, Inc. Wolfram Research retains such copyright ownership, but does hereby grant the Publisher worldwide rights to use and distribute this software as it pertains to this work. Permission to use and reproduce copies of any part of this software must be obtained from the creators by contacting the Publisher in writing at TELOS, The Electronic Library of Science, 3600 Pruneridge Avenue, Suite 200, Santa Clara, CA 95051.

The use of general descriptive names, trademarks, etc., in this publication, even if the former are not especially identified, is not to be taken as a sign that such names, as understood by the Trade Marks and Merchandise Marks Act, may accordingly be used by anyone. Where those designations appear in this work and Springer-Verlag was aware of a trademark claim, the designations follow the capitalization style used by the manufacturer.

MathReader is a trademark and *Mathematica* is a registered trademark of Wolfram Research, Inc., Macintosh is a registered trademark of Apple Computer, Inc., NeXT and NeXTSTEP are trademarks of NeXT Computer, Inc., Windows is a trademark of Microsoft Corporation, Unix is a registered trademark of AT&T. SPARC and SunOS are registered trademarks of Sun Microsystems, Inc.

Pages prepared from the author's *Mathematica* notebook and converted to LaTeX.
Printed in the United States of America.

9 8 7 6 5 4 3 2 1

ISBN 0-387-14222-3

TELOS, The Electronic Library of Science, is an imprint of Springer-Verlag New York with publishing facilities in Santa Clara, California. Its publishing program encompasses the natural and physical sciences, computer science, economics, mathematics, and engineering. All TELOS publications have a computational orientation to them, as TELOS' primary publishing strategy is to wed the traditional print medium with the emerging new electronic media in order to provide the reader with a truly interactive multimedia information environment. To achieve this, every TELOS publication delivered on paper has an associated electronic component. This can take the form of book/diskette combinations, book/CD-ROM packages, books delivered via networks, electronic journals, newsletters, plus a multitude of other exciting possibilities. Since TELOS is not committed to any one technology, any delivery medium can be considered.

The range of TELOS publications extends from research level reference works through textbook materials for the higher education audience, practical handbooks for working professionals, as well as more broadly accessible science, computer science, and high technology trade publications. Many TELOS publications are interdisciplinary in nature, and most are targeted for the individual buyer, which dictates that TELOS publications be priced accordingly.

Of the numerous definitions of the Greek word "telos," the one most representative of our publishing philosophy is "to turn," or "turning point." We perceive the establishment of the TELOS publishing program to be a significant step towards attaining a new plateau of high quality information packaging and dissemination in the interactive learning environment of the future. TELOS welcomes you to join us in the exploration and development of this frontier as a reader and user, an author, editor, consultant, strategic partner, or in whatever other capacity might be appropriate.

TELOS, The Electronic Library of Science
Springer-Verlag Publishers
3600 Pruneridge Avenue, Suite 200
Santa Clara, CA 95051

TELOS Diskettes

Unless otherwise designated, computer diskettes packaged with TELOS publications are 3.5" high-density DOS-formatted diskettes. They may be read by any IBM-compatible computer running DOS or Windows. They may also be read by computers running NEXTSTEP, by most UNIX machines, and by Macintosh computers using a file exchange utility.

In those cases where the diskettes require the availability of specific software programs in order to run them, or to take full advantage of their capabilities, then the specific requirements regarding these software packages will be indicated.

TELOS CD-ROM Discs

For buyers of TELOS publications containing CD-ROM discs, or in those cases where the product is a stand-alone CD-ROM, it it always indicated on which specific platform, or platforms, the disc is designed to run. For example, Macintosh only; Windows only; cross-platform, and so forth.

TELOSpub.com (Online)

Interact with TELOS online via the Internet by setting your World-Wide-Web browser to the URL: http://www.telospub.com.

The TELOS Web site features new product informatin and updates, an online catalog and ordering, samples from our publications, information about TELOS, data-files related to and enhancements of our products, and a broad selection of other unique features. Presented in hypertext format with rich graphics, it's your best way to discover what's new at TELOS.

TELOS also maintains these additional Internet resources:

> gopher://gopher.telospub.com
>
> ftp://ftp.telospub.com

For up-to-date information regarding TELOS online services, send the one-line e-mail message:

> send info to: info@TELOSpub.com.

Illustrated Mathematics

Table of Contents

1 **What Is *Illustrated Mathematics*?** 7
  – Collection of Visualizations   7
  – Programs for Your Own Examples   8

2 **Topics of *Illustrated Mathematics***   9
  – Theory of Functions   9
  – Sequences and Series   11
  – Derivatives   12
  – Analysis of Functions   13
  – Integration   14
  – Differential Equations   14
  – Conic Sections   17
  – Complex Functions   17
  – Linear Maps   19
  – Conformal Maps   20
  – Cycloids and Related Curves   20
  – Figures of Revolution   21
  – Polyhedra   22
  – Icosahedra   23
  – Minimal Surfaces   24
  – Iterated Functions   24

3 **Directories/Folders and Files on the CD-ROM**   25
  – Collection of Visualizations   25
  – Programs for Your Own Examples   26
  – Versions   27

4 **Hardware and Software Prerequisites**   28
  – Collection of Visualizations   28
  – Programs for Your Own Examples   28
  – Color Monitor   29

5 **Installation Instructions**   30
  – General Remarks   30
  – Macintosh   31
  – Windows   32

- NeXTSTEP 33
- Unix 35

6 Notebook Documents 38
- Groups: The Structure of Notebooks 38
- Graphics and Animations 39
- Window Size and Display Size 40
- Printing 41
- Copying Graphics into Other Documents 42
- Run-Time Problems 42

7 Using the Programs 43
- *Mathematica's* User Interface 43
- Loading Programs: Needs 43
- Documentation 44

8 General Remarks about *Mathematica* 45
- Syntax 45
- Options 47

9 Graphics with *Mathematica* 49
- Graphic Commands 49
- Graphic Objects 53
- Options 54

10 Selected Publications about *Mathematica* 62
- Introductory 62
- Handbooks/Reference 62
- Programming 62
- Applications 63
- Teaching 63
- Periodicals 63

11 Further Development of *Illustrated Mathematics* 64

Illustrated Mathematics

Visualization of Mathematical Objects
with *Mathematica*

Oliver Gloor   Beatrice Amrhein   Roman E. Maeder

Illustrated Mathematics is a comprehensive collection of graphics and animations for various topics in mathematics on a CD-ROM. The images can be used without any additional software.

Further, programs for the design of new examples are included. The programs are written in *Mathematica's* programming language, which means that *Mathematica* is necessary for their use. Prior knowledge of *Mathematica* is not required. Examples can be created by modifying parameters in existing examples.

Illustrated Mathematics is designed for teachers and students at high schools and undergraduate colleges.

The software can be used on Macintosh computers and under Windows (without any additional software), and on other computers (together with *Mathematica* or *MathReader*). The CD-ROM contains fully worked out collections for Macintosh and Windows. Versions for other computers (NeXT and Unix) are included in compressed form.

The CD-ROM is formatted in dual-mode, that is, it looks like an ordinary (HFS) volume on the Macintosh. On other computers, the ISO 9660 part can be accessed.

Colophon

Author addresses:

Oliver Gloor
Universität Tübingen
WSI Sand 13
D-72076 Tübingen
`gloor@informatik.uni-tuebingen.de`

Beatrice Amrhein
Universität Tübingen
WSI Sand 13
D-72076 Tübingen
`amrhein@informatik.uni-tuebingen.de`

Roman Maeder
ETH Zürich
ETH Zentrum IFW
CH-8092 Zürich
`maeder@inf.ethz.ch`

In one form or another, the following persons have helped us with this project: Urs Kirchgraber, Werner Hartmann, Hermann-Josef Biner, Stéphane Collart, Markus Furter, and Albrecht Plümicke. We thank all of them.

We thank Allan Wylde, Paul Wellin, and the staff at TELOS for their cooperation in publishing *Illustrated Mathematics*.

Zürich, March 1995   Oliver Gloor,
                       Beatrice Amrhein,
                       Roman E. Maeder

1 What Is *Illustrated Mathematics*?

Visual presentations aid in understanding complicated objects. Understanding abstract constructions frequently encountered in mathematics is improved with visual representations and graphics. Such visualizations are also important for students who have difficulty grasping abstract objects.

If visualization had been our only goal, we could have published this collection as a printed book instead of as a CD-ROM. CD-ROMs can store much more information, however. Besides, we can animate sequences of graphics and the computer allows direct interaction with the material. For example, it is feasible to create additional images during class and respond to students' questions. This capability improves the quality of instruction.

Collection of Visualizations

The graphics and animations on the CD-ROM can be used directly for teaching. No additional software is necessary (on Macintosh and under Windows).

Projection

Examples from this collection can be projected on a screen during class (for example, with an LCD projection panel connected to your computer).

Hardcopy

If no computer-compatible projection equipment is available, the examples can be printed for class notes or to create slides.

Part of Documents

The graphics and animations are part of hierarchically structured documents (*Mathematica* notebooks). This organization allows fast access to particular examples and enables us to add explanatory text.

For Class Notes

Either directly on the computer or printed out, the graphics can be included in your documents and used as part of class notes. To help with this use of the graphics, many graphics are stored on the CD-ROM outside of their notebooks, in a graphics standard format.

Color

Because many graphics use color, a color monitor is recommended. However, colors have been chosen in a way that makes presentations with a monochrome monitor possible.

Included: *MathReader*

The *MathReader* program included on the CD-ROM allows you to read all documents on this CD-ROM, including animations.

Programs for Your Own Examples

Mathematica Is Necessary

The programs are based on *Mathematica*. For their use, *Mathematica* is necessary. In fact, the collection of examples was created with these programs.

Prior Knowledge of *Mathematica* Is Not Necessary

Some experience with *Mathematica* is an advantage, but not required for the use of the *Illustrated Mathematics* programs. Simple adjustment of parameters in the examples allows their modification. Each program (package) is accompanied by a manual and many examples of program use.

Creation of Teaching Material

The programs allow you to create your own teaching material in the form of *Mathematica* notebooks, the form also used for this collection.

2 Topics of *Illustrated Mathematics*

Each part of *Illustrated Mathematics* is listed here with an example. All examples have been produced exclusively with the programs that are part of *Illustrated Mathematics*.

Since this booklet cannot show true animations, one or several selected graphics are shown.

Additional types of visualizations can be found in the documents that constitute the collection and in the description of the commands.

Theory of Functions

Examples can be found in the **Collection** notebooks **Functions** and **TrigFunctions**. Notices to the programs are in the **Manual** notebooks **Analysis** and **TrigFunctions**.

Polynomials and Rational Functions

The influence of the parameter **a** in the cubic polynomial $x^3 + a\,x$ is shown in the sequence of images. The value of **a** is changed from **-1** to **1**.

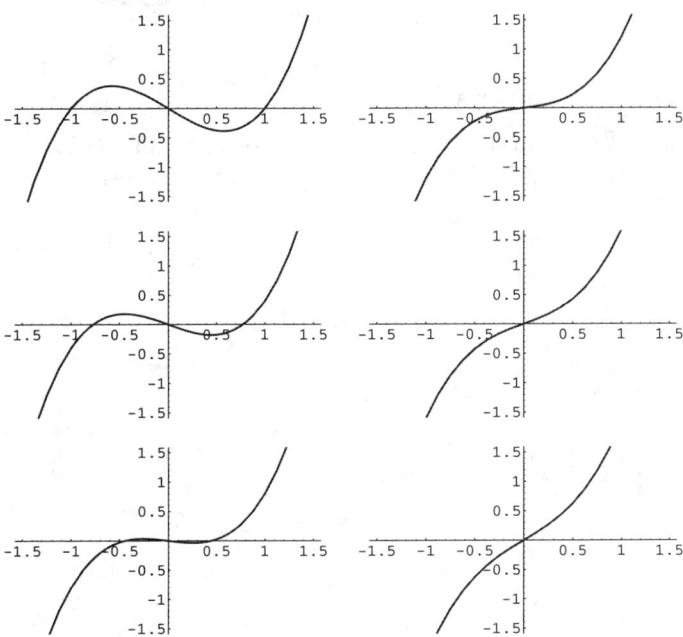

Trigonometric Functions

The construction of the sine function with the help of a triangle with hypotenuse 1.

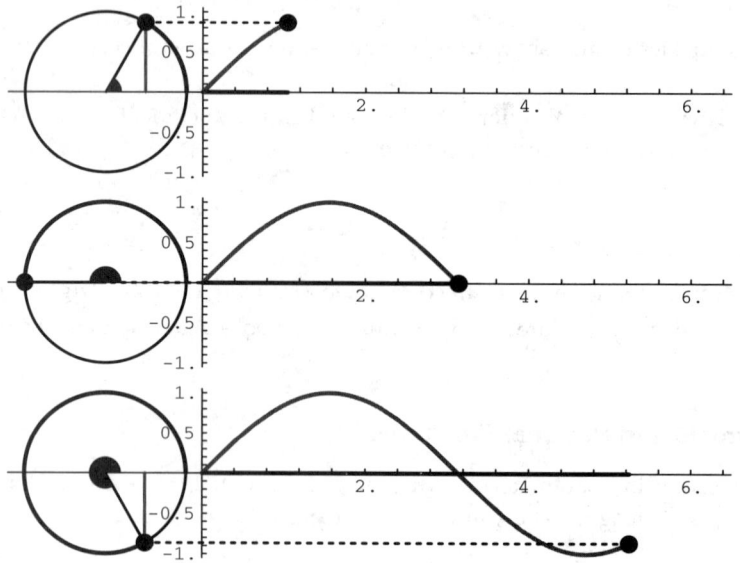

Exponential and Logarithm

Two logarithms, the natural logarithm (dark) and the logarithm to base 10 (light), as well as two exponentials, **e^{-x}** (light) and **10^{-x}** (dark) are shown.

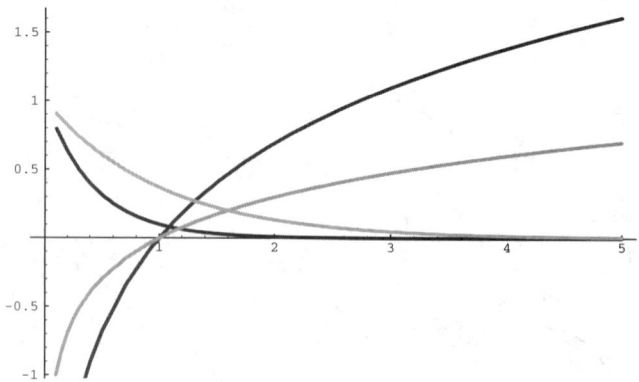

Sequences and Series

The sequence **1/n** (dark) converges more slowly toward zero than **1/n^2** (light).

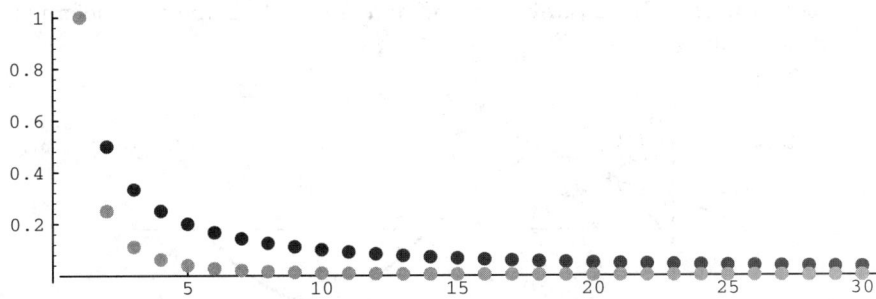

The alternating harmonic series converges toward **log 2**.

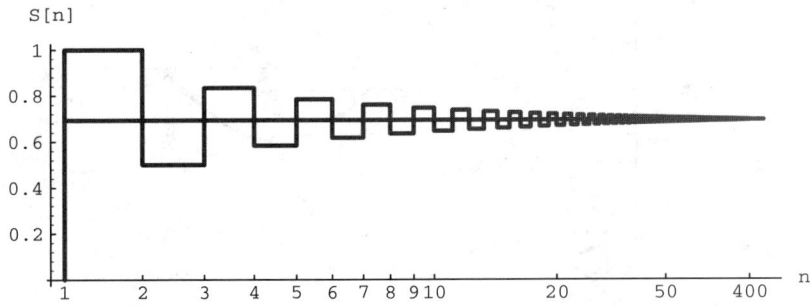

Examples can be found in the **Collection** notebooks **Sequences** and **Series**. Notices to the programs are in the **Manual** notebooks **Sequences** and **Series**.

Derivatives

This animation shows how the derivative of a function is constructed. The tangent through the point running along the curve is constructed and translated to the left. There, its slope can be read off and is used to graphically determine the first derivative.

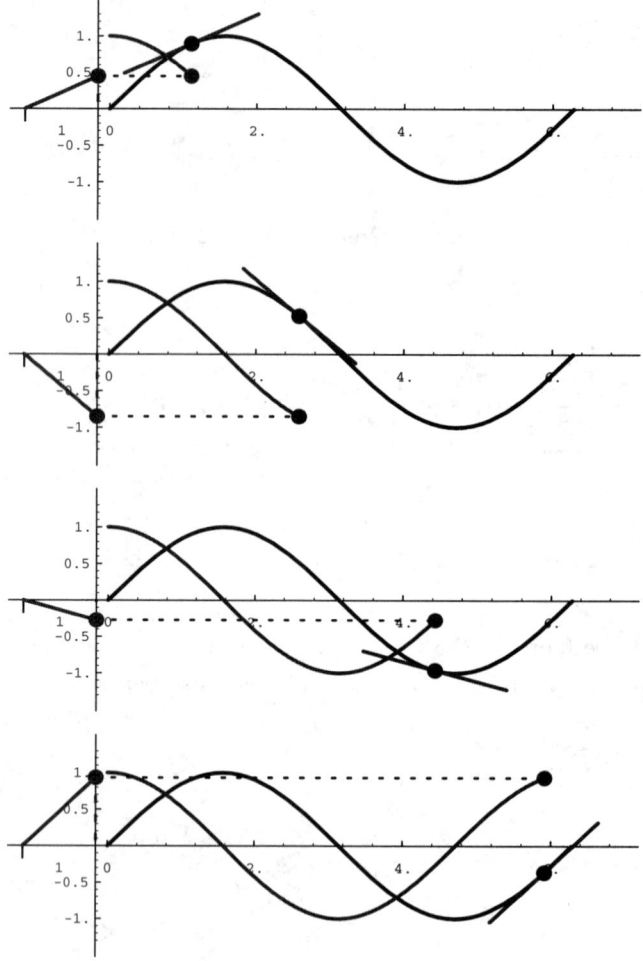

Examples can be found in the **Collection** notebooks **Derivatives** and **Differential**. Notices to the programs are in the **Manual** notebook **Differentiation**.

Analysis of Functions

The popular discussion topic of functions can be supported by computer. Here zeros, extremal values, and points of inflection are computed and shown graphically.
Analysis of the rational function $f(x) = x(x-2)(x+2)/(x2+1)$:

Function: $\dfrac{x(-4 + x^2)}{1 + x^2}$

Symmetry with respect to the origin

Zeros: {-2, 0, 2}

Minima: {(0.728786,-1.65111)}

Maxima: {(-0.728786,1.65111)}

Points of Inflection: {(-1.73205,0.433013), (0,0),
   (1.73205,-0.433013)}

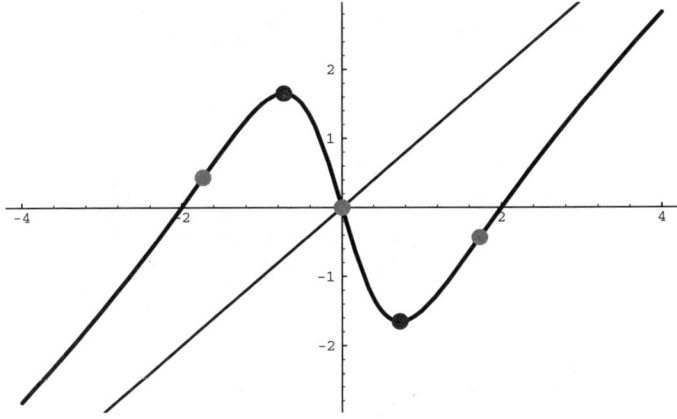

Examples can be found in the **Collection** notebook **Analysis**.
Notices to the programs are in the **Manual** notebook **Analysis**.

Integration

The construction of the integral is demonstrated in this animation. The area under the curve is displayed on the left in the form of a rectangle with base 1. The height of this rectangle is the value of the integral. In the animation, the upper limit of the integral is increased from left to right.

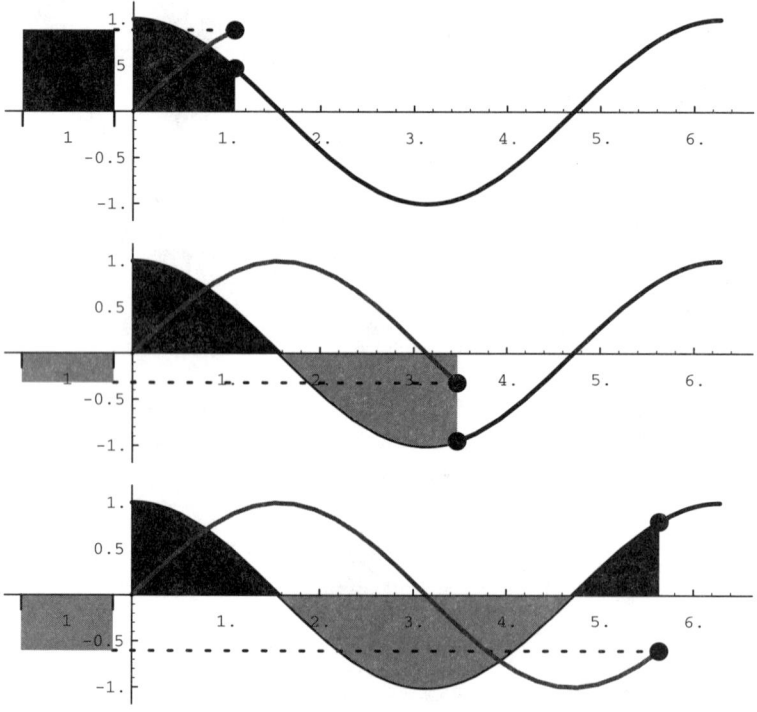

Examples can be found in the **Collection** notebook **Integration**.
Notices to the programs are in the **Manual** notebook **Integration**.

Differential Equations

Examples can be found in the **Collection** notebooks **ODEs, ODEApplications,** and **ODEAnimations**.
Notices to the programs are in the **Manual** notebook **ODEs**.

First-Order Differential Equations

A linear differential equation of first order specifies a direction for each point in a region. It is the tangent direction of the integral curve through this point. This is the direction field for the equation **y' = -x/y**.

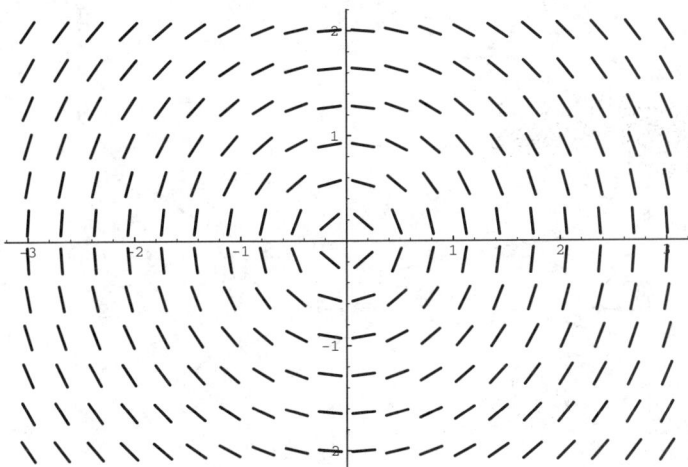

These are the solutions of **y y' + x = 0**:

Solutions: {-Sqrt[-x^2 + C[1]], Sqrt[-x^2 + C[1]]}

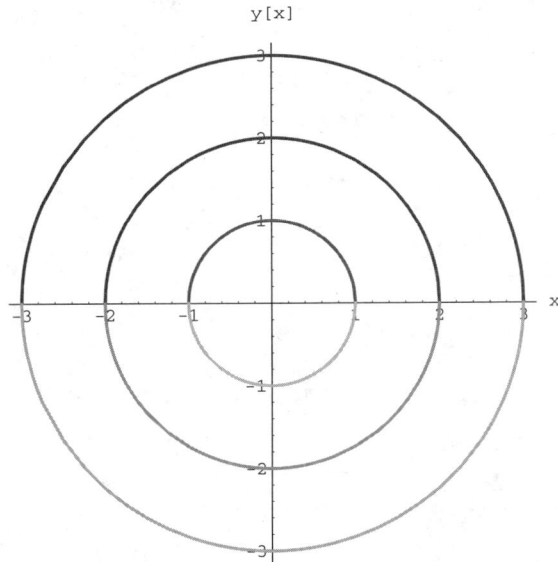

Second-Order Differential Equations

We show the solutions of the differential equation $y'' - y' - y = 0$ with boundary conditions $y(1) = a$ and $y'(1) = b$ for a, b in $\{-1, 0, 1\}$.

Solutions: $\{E^{((1 - \text{Sqrt}[5])\, x)/2}\, C[1] + E^{((1 + \text{Sqrt}[5])\, x)/2}\, C[2]\}$

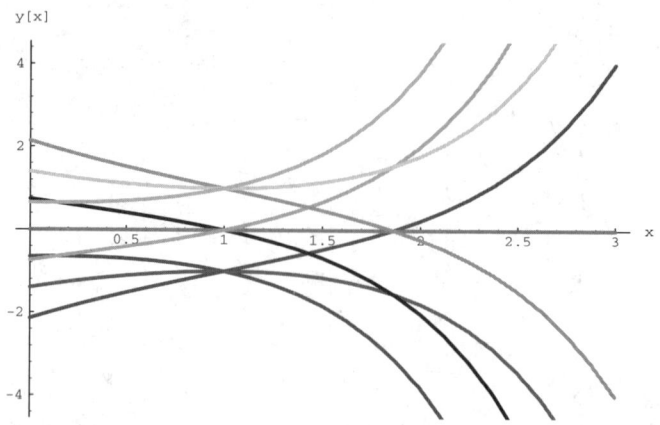

This is a graphic with solutions of $\cos(3/2\ t) = x'' + x$ in phase space.

Solutions: $\{C[2]\, \text{Cos}[t] - \dfrac{4\, \text{Cos}[\frac{3\,t}{2}]}{5} - C[1]\, \text{Sin}[t]\}$

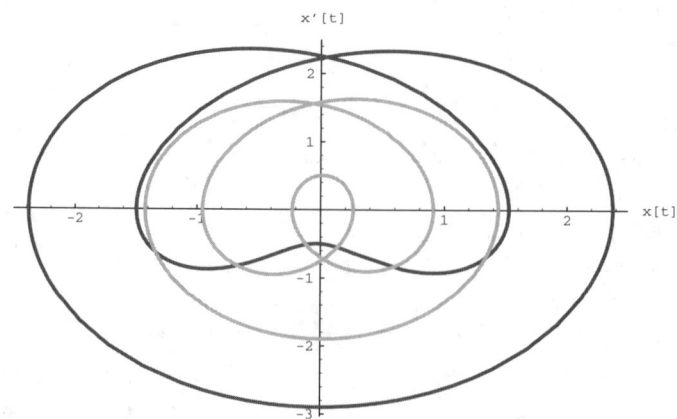

Conic Sections

The zero manifold of a quadratic polynomial in two variables forms a conic section. The characteristics of this (possibly degenerate) conic (its center, apices, focal points, and asymptotes) are computed and shown in the graphic.

Parabolas, ellipses, and hyperbolas as the locus of points that satisfy certain distance constraints can be visualized by an animation.

Reflection properties of these conic sections are also illustrated.

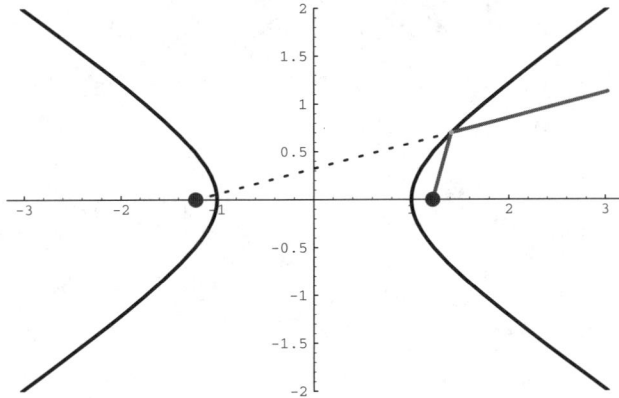

Examples can be found in the **Collection** notebooks **ConicSections**, **Reflections**, and **Classification**.

Notices to the programs are in the **Manual** notebook **ConicSections**.

Complex Functions

A complex-valued function requires four real dimensions for visualization (two each for domain and range). This can be achieved in three dimensions by using color as the fourth dimension: the horizontal plane represents the domain (real and imaginary part), the vertical axis shows the absolute value of the function, the color shows the argument according to this key.

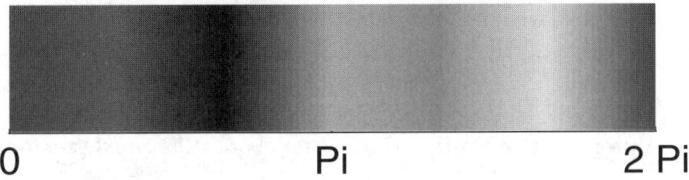

This is the identity function **f(z) = z**:

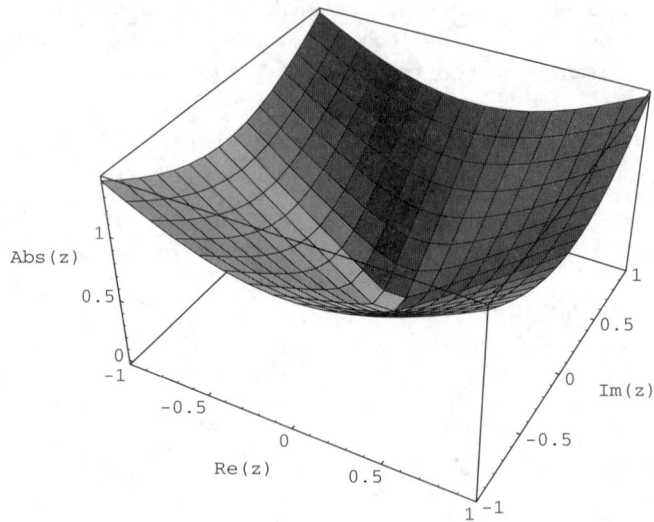

The function **f(z) = 1/z^2** has a double pole at **z = 0**.

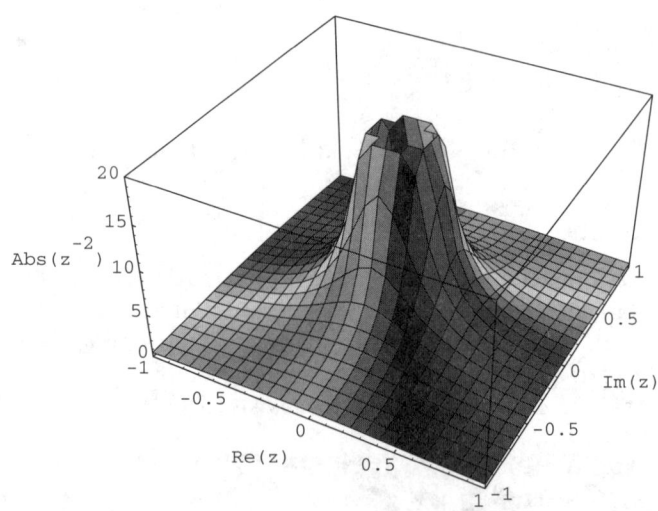

Examples can be found in the **Collection** notebook **ComplexFunctions**.
Notices to the programs are in the **Manual** notebook **ComplexFunctions**.

Linear Maps

This is the effect of a linear mapping on Cartesian parameter lines.

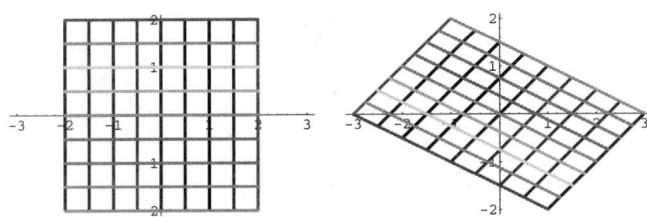

If a two-dimensional linear map has two real-valued eigenvectors, any point can be written in terms of these and each component can be mapped separately. In an animation, this process can be performed for each point of the unit circle.

```
M =              1.83333      0.166667
                 0.333333     1.66667
Eigenvalues:  {2., 1.5}
Eigenvectors: {{0.707107, 0.707107}, {-0.447214, 0.894427}}
```

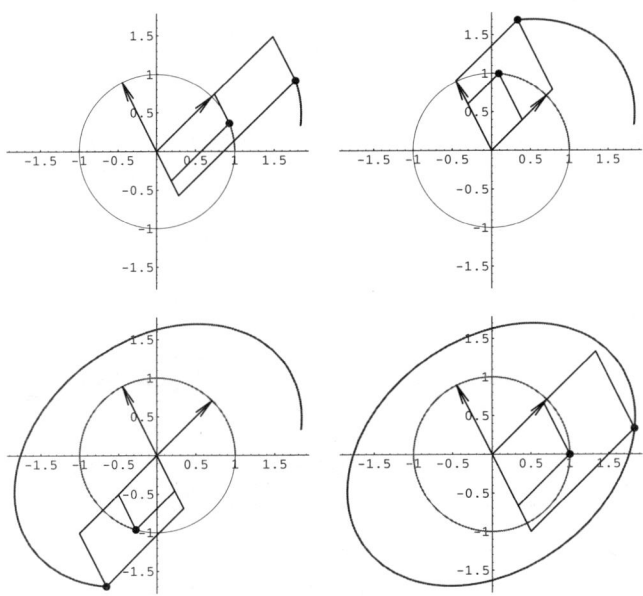

Examples can be found in the **Collection** notebook **LinearMaps**.
Notices to the programs are in the **Manual** notebook **LinearMaps**.

Conformal Maps

Conformal maps can be visualized using the images of coordinate lines. Here are the polar coordinate lines under the Möbius transform **f(z) = (2z-I)/(z-1)**:

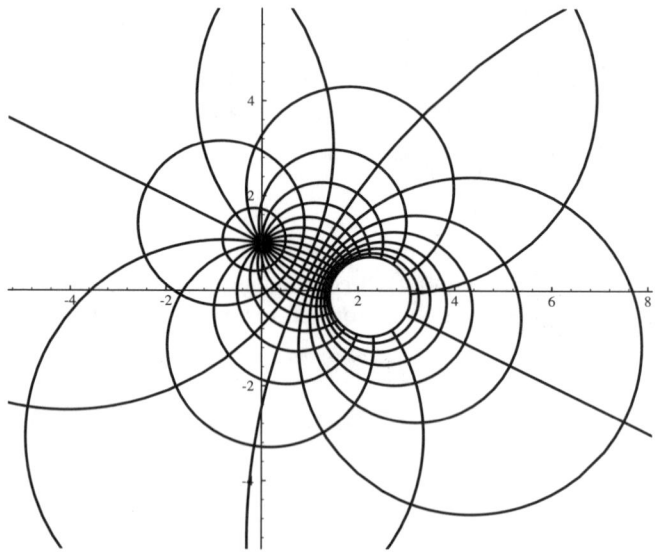

Examples can be found in the **Collection** notebook **Conformal**.
Notices to the programs are in the **Manual** notebook **ComplexMap**.

Cycloids and Related Curves

This is a cycloid.

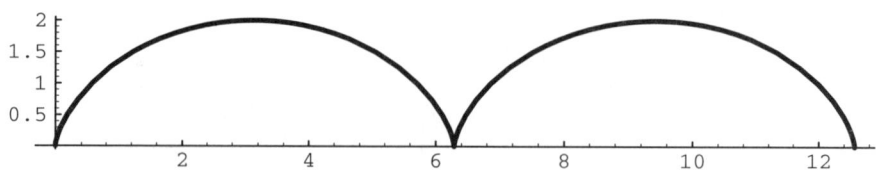

The curve traced by a point with a distance of **0.8** from the center of a circle (radius **1**) which rolls in a circle with radius **3/2** is a curtate hypocycloid.

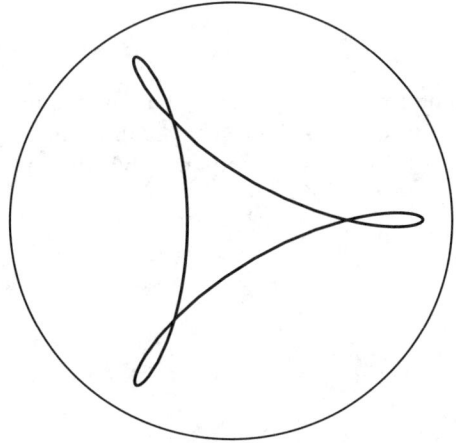

Examples can be found in the **Collection** notebooks **RollingCircles**, **Cycloids**, **Hypocycloids**, and **Epicycloids**.
Notices to the programs are in the **Manual** notebook **RollingCircles**.

Figures of Revolution

Cylinder, sphere, torus, Möbius strip, double helix, and so on, are surfaces of revolution. The last two of them are twisted.

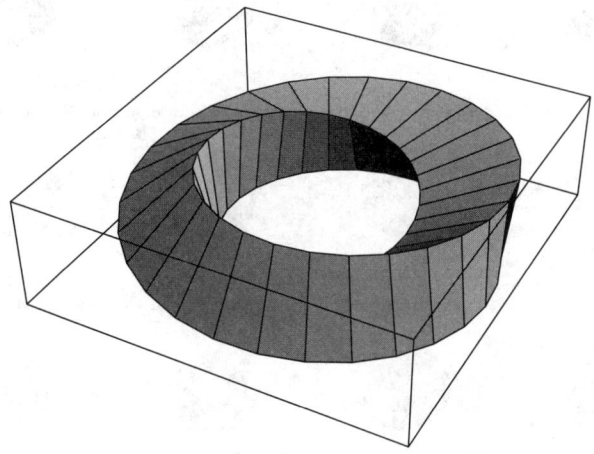

The creation process of such surfaces can be shown by an animation.
Examples can be found in the **Collection** notebook **Revolution**.
Notices to the programs are in the **Manual** notebook **Revolution**.

Polyhedra

There are exactly 75 uniform polyhedra, as well as two infinite families of prisms and antiprisms. This picture shows the 20 convex uniform polyhedra, that is, the five Platonic solids, the thirteen Archimedian polyhedra, as well as one prism and one antiprism.

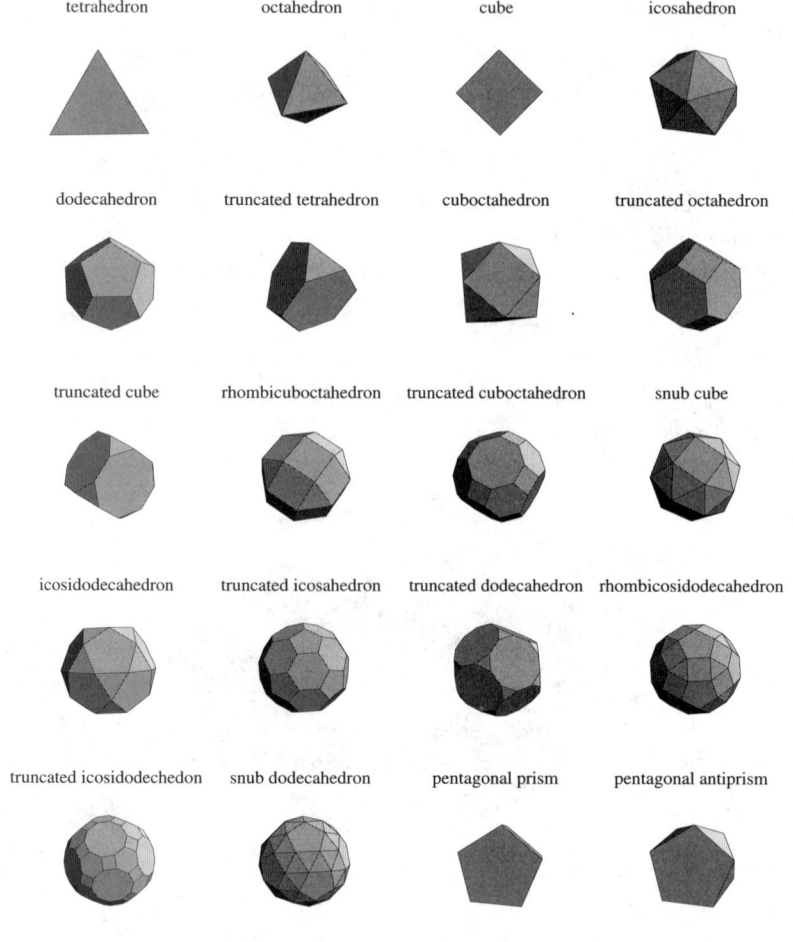

Examples can be found in the **Collection** notebooks **Polyhedra** and **PolyPictures**. Notices to the programs are in the **Manual** notebook **Polyhedra**.

Icosahedra

There are 59 stellations of the icosahedron. Here are the original icosahedron, the compound of five octahedra, the commpound of five tetrahedra, and the great icosahedron.

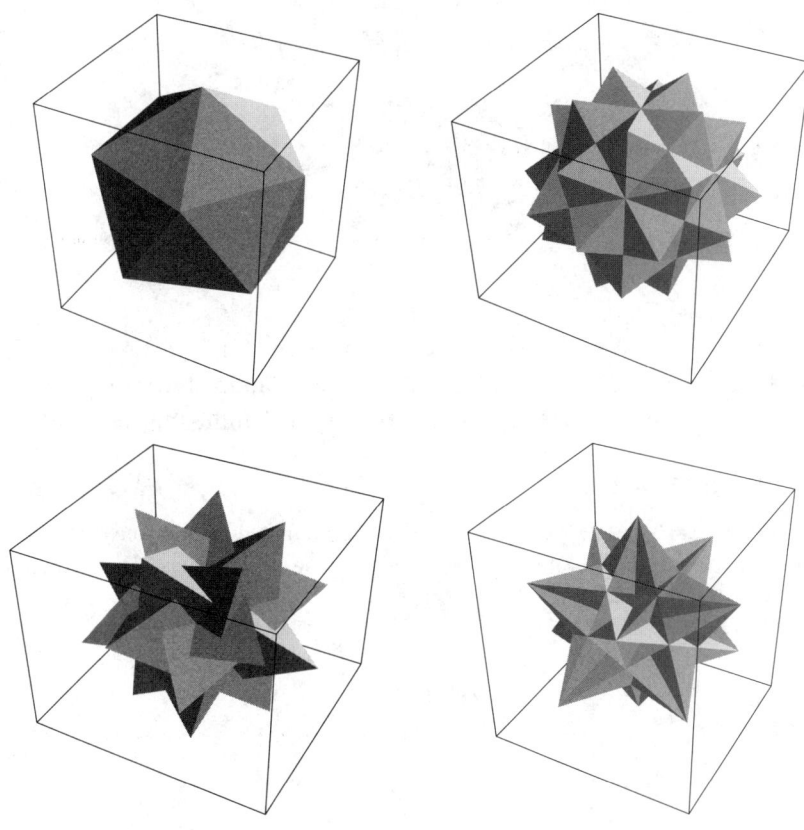

Examples can be found in the **Collection** notebook **Icosahedra**.
Notices to the programs are in the **Manual** notebook **Icosahedra**.

Minimal Surfaces

Minimal surfaces are important in physics and engineering. This picture shows Enneper's surface with polar parameter lines.

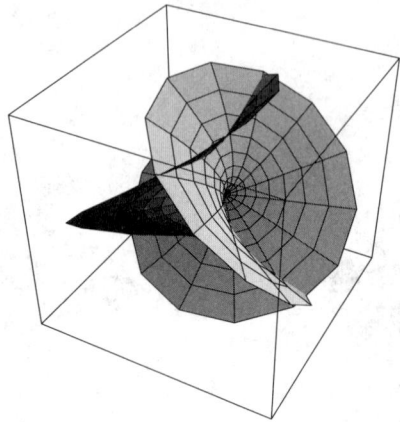

Examples can be found in the **Collection** notebook **MinimalSurfaces**.
Notices to the programs are in the **Manual** notebook **MinimalSurfaces**.

Iterated Functions

The function `f(x) = 4x(1-x)` is one of the simplest examples of chaotic behavior. If the function is iterated repeatedly, close initial values are distributed arbitrarily in the interval `[0,1]`.

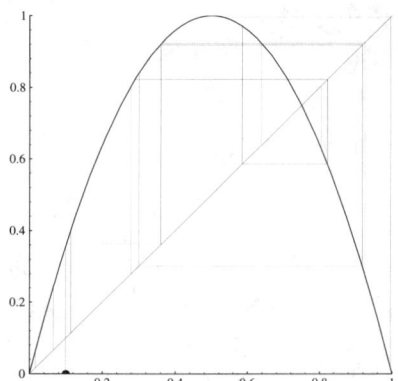

Examples can be found in the **Collection** notebook **Chaos**.
Notices to the programs are in the **Manual** notebook **Chaos**.

3 Directories/Folders and Files on the CD-ROM

This section briefly describes the directories/folders and files, which are grouped by purpose.

Windows version: Filenames and directory names have been truncated after eight characters, for example, **COLLECTI** instead of **Collection**. Most notebooks are accompanied by another file with an **.MB** extension that contains binary information for *MathReader*.

NeXT/Unix version: Notebooks and programs were compressed into a single tar file. This archive is named **NEXT_TAR.Z** and **UNIX_TAR.Z**, respectively. The installation instructions describe how these archives can be unpacked.

Collection of Visualizations

Collection

The **Collection** directory/folder contains all graphics and animations as notebooks. No further installation is necessary to use them. Windows version: According to the installation instructions two files are to be copied into your **WINDOWS** directory.

Since *MathReader* does not allow users to copy and paste graphics into other applications, many graphics have been stored as individual files in the **Images** subdirectory. More information about them can be found in the **Hints.ma** document in the **Images** directory.

Macintosh: Some notebooks are rather large. Additional copies of these have been created, with their graphics converted into a more compact format. These versions should be used on machines with less main memory. The **PICT** folder contains those notebooks whose graphics were converted into PICT format. Those notebooks that were still over 2MB after this conversion have been put into the **Bitmap** folder, using bitmap graphics.

Table of Contents

The **Contents.ma** file contains an overview of the visualization and program topics. It is reprinted in the booklet accompanying the CD-ROM.

Using Notebooks and *MathReader*

Macintosh and Windows version: The **MathReader** directory/folder contains the *MathReader* program from Wolfram Research, Inc. If *Mathematica* is not installed on your computer, you can use this program to read notebooks.

Versions of *MathReader* for other computers may be available on *MathSource* (see Installation Instructions).

The **Notebooks.ma** file in the **Introduction** directory/folder contains an introduction to the use of *MathReader*. It is reprinted in the booklet accompanying the CD-ROM.

About *Illustrated Mathematics*

A description of *Illustrated Mathematics* is in the **About_IlluMath.ma** (Windows: **ABOUT_IM.MA**) file and reprinted in the booklet accompanying the CD-ROM.

Programs for Your Own Examples

The following directories/folders and files pertain to the use of the programs.

Programs and Their Descriptions

The **Visual** directory contains the programs. They should be copied to a hard disk according to the installation instructions. Alternatively, you can tell *Mathematica* the pathname where they can be found.

Each program, or more precisely, each package, is accompanied by a detailed description in the **Manuals** directory which includes many examples.

Installation Instructions

For best use, the packages should be copied to a hard disk. The **Installation.ma** file contains instructions for installing the software on various computers. Installation is necessary only if you want to create your own examples. These installation instructions are also printed in the accompanying booklet.

Computer-dependent Settings

The **Settings.ma** notebook contains settings for configuring *Illustrated Mathematics* for various computers (for example, color or monochrome monitors). It is needed only if you want to create your own examples.

Introduction into Notebooks and *Mathematica*

The **Introduction** directory/folder contains introductory files. **Notebooks.ma** describes how to look at notebooks. Notebooks can be displayed either with *MathReader* or with *Mathematica*. **Programs.ma** describes the use of the programs on the CD-ROM. **Mathematica.ma** explains the basic principles of *Mathematica* and describes how to work with *Mathematica* graphics. It also contains a summary of the built-in graphics capabilities of *Mathematica*.

Versions

The CD-ROM is formatted in dual-mode, HFS/ISO 9660. If read on a Macintosh, only the files for the Macintosh will be visible. If read on another computer (Windows or Unix), the documents for Windows and the two files **NEXT_TAR.Z** and **UNIX_TAR.Z** will be visible. The installation of these versions is described in **Installation.ma**.

4 Hardware and Software Prerequisites

To read the CD-ROM, a CD drive with any necessary operating system extensions is required. The CD-ROM is formatted in dual-mode, HFS/ISO 9660, and can be read on most computers.

Collection of Visualizations

Macintosh / Windows

All graphics and animations can be read or printed without any additional software on a Macintosh (System 7) or under Windows (Version 3.0 or later), using the enclosed program *MathReader*. At least 4MB of RAM are required.

NeXT

No notebook reader is included on the CD-ROM; therefore, *Mathematica* is required even to read or print the notebooks. All files have been compressed into a single archive (see Installation Instructions).

Unix/X11

No notebook reader is included on the CD-ROM. Such versions of *MathReader* may be available on *MathSource* (see Installation Instructions). Otherwise, a version of *Mathematica* with the notebook front end is required even to read or print the notebooks. For computers supporting X11, *Mathematica* Version 2.2 provides such a front end. It can be ordered from your *Mathematica* dealer, if necessary.

You can open the notebooks of the Windows version with any notebook reader. However, we recommend you use the proper version for Unix. All files have been compressed into a single archive (see Installation Instructions).

Programs for Your Own Examples

Mathematica Version 2.0 and enough memory (at least 5MB) for its use are required to run the programs of *Illustrated Mathematics*. No CD drive is necessary to run the programs.

The programs are machine-independent.

All the examples were computed with *Mathematica* Version 2.2 on Macintosh, NeXT, Windows, or Sun/SPARC systems.

Macintosh / Windows / NeXT

Mathematica allows you to read the notebooks. Programs are easiest to use by modifying an example in one of the notebooks from either the **Collection** or **Manual** directories/folders.

Other Computers

The programs can be used with *Mathematica* on any computer. To read the notebooks, however, you need the notebook front end. Such a front end is available for the X11 window system starting with Version 2.2 of *Mathematica*.

Using the front end you can modify an example in one of the notebooks of *Illustrated Mathematics* from either the **Collection** or **Manual** directories.

Color Monitor

A color monitor is recommended, but not required.

5 Installation Instructions

The CD-ROM is formatted in dual-mode, that is, it looks like an ordinary (HFS) volume on the Macintosh. On other computers, the ISO 9660 part can be accessed. It contains all files for the Windows version and the two files **NEXT_TAR.Z** and **UNIX_TAR.Z**, containing the files for NeXT and Unix computers.

The accompanying booklet and the **About_IlluMath.ma** notebook on the CD-ROM contain information about hardware and software requirements, as well as information about the two parts of *Illustrated Mathematics:*
– Collection of Visualizations
– Programs for Your Own Graphics

Information on the use of the software is printed in the booklet and can be found in the **Introduction** directory/folder.

General Remarks

Problems Opening Collection Notebooks

If you run out of memory looking at a notebook, you may try to close and reopen the notebook.

Macintosh: Note that many notebooks in the collection have been stored with converted graphics as well. This format allows you to open them with a smaller amount of memory (RAM). Furthermore, in the Macintosh section of the Installation Instructions, you find instructions on how to give *MathReader* more memory.

Windows: To save space, some notebooks have been stored without the binary information. If such a notebook is opened, the missing data are restored automatically, including bitmaps for graphics displayed on the screen. When giving a demonstration, you may want to open the notebook beforehand to allow the rendering of graphics to complete before you use the notebook. If you use *Mathematica*, the binary information will be saved when you save the notebook.

Currently, CD-ROM drives are much slower than hard disks. You may want to copy frequently used notebooks onto a hard disk.

The first iteration through an animation is often slower, since all data has to be read first.

Use of *Mathematica* 2.0 instead of 2.2

The programs can be used with *Mathematica* Version 2.0. When reading the prepared notebooks, graphics may not be visible, however, since Version 2.0 cannot display all graphics created with Version 2.2. You can either compute the graphics anew or you may want to look at the notebooks with the *MathReader* program, enclosed on the CD-ROM.

Memory Problems during Computation of Graphics

For most animations you can set the number of frames to render with the option **Pictures -> k**. This allows you to render only a few frames during development. You can change the global default value for all animations in the **Settings.ma** notebook.

The option **PlotPoints -> k** allows you to modify the number of points rendered. Especially when testing three-dimensional graphics, it is worthwhile to use a smaller number of points.

Macintosh

Collection of Visualizations

No installation is necessary. You can double-click the desired documents (notebooks) on the CD-ROM. *MathReader*, the program used to display notebooks, is started automatically.

There are some ways to improve response time, which are especially valuable for larger notebooks or animations.

To use *MathReader* for larger notebooks, copy it onto a hard disk. Now, you can give it more memory: select it and choose **Get Info** in the **File** menu. By default, *MathReader* uses 2560kB, but you can choose to increase this number, depending on your configuration. Using virtual memory (under System 7), you can work with large notebooks even with small amounts of RAM.

If you use a particular notebook frequently, you should copy it onto a hard disk to improve access time.

Programs for Your Own Graphics

To use the programs for the creation of your own examples, we recommend you copy the *Mathematica* packages in the **Visual** folder onto a hard disk.

Mathematica must find the **Visual** directory and the **Settings.ma** notebook. To allow the *Mathematica* kernel to find these programs, you must either install the programs in a place searched by *Mathematica*, or you must tell the kernel where to find them.

You should copy the **Visual** folder and the **Settings.ma** notebook into the *Mathematica* **Packages** folder. These two objects (**Visual** and **Settings.ma**) will then be at the same level as the **Algebra** folder or the **init.m** package, for example.

If you install the **Visual** folder and **Settings.ma** in a different place or choose to leave them on the CD-ROM, or if you don't have write access to the *Mathematica* folder, you must tell *Mathematica* where to find the packages before trying to read them.

If you want to use the packages directly from the CD-ROM, you can tell *Mathematica* where to find them by this command.

    $Path = Join[$Path, {"Illustrated Mathematics"}];

If you copy **Visual** and **Settings.ma** into the subfolder **1995** of **Pictures**, where **Pictures** is a subfolder of **Programs**, for example, you should use

    $Path = Join[$Path, {"Programs:Pictures:1995"}];

To have *Mathematica* perform this command automatically at startup, you should write this command into the **init.m** notebook in the *Mathematica* **Packages** folder.

Windows

Collection of Visualizations

No installation is necessary. You can start *MathReader* (**MATHREAD.EXE** in the directory **MATHREAD**) directly from the CD-ROM and then use it to open the desired notebook.

MathReader requires a file **MATHREAD.INI** located in the **WINDOWS** directory of your hard disk. If you haven't installed *MathReader*, this file may not be there. In this case, *MathReader* will ask you to locate it. It can be found in the **MATHREAD** directory on the CD-ROM. We recommend that you copy *CD-ROM***:\MATHREAD\MATHREAD.INI** into your Windows directory to avoid this inconvenience. You may also want to copy *CD-ROM***:\MATHREAD\CTL3D.DLL** to the **SYSTEM** subdirectory of your **WINDOWS** directory.

There are some ways to improve response time, which are especially valuable for larger notebooks or animations.

To speed up *MathReader*, you should install it on your hard disk. Follow the instructions in the **INSTALL.TXT** file in the **MATHREAD** directory on the CD-ROM to install *MathReader* on your hard disk.

If you install *MathReader*, a default **MATHREAD.INI** file will be copied into your **WINDOWS** directory. We recommend you replace it with our version in the **MATHREAD** directory of the CD-ROM (*after* the installation of *MathReader*).

If you use a particular notebook frequently, you should copy it onto a hard disk to improve access time.

If you have trouble opening a notebook (e.g., if you use another version of *Mathematica*), you may try to open it with the **No Binary** box checked.

Programs for Your Own Graphics

To use our programs for the creation of your own examples, we recommend you copy the *Mathematica* packages in the **VISUAL** directory onto a hard disk.

Mathematica must find the **Visual** directory and the **Settings.ma** notebook. To allow the *Mathematica* kernel to find these programs, you must either install the programs in a place searched by *Mathematica*, or you must tell the kernel where to find them.

You should copy the **VISUAL** directory and the **SETTINGS.MA** notebook into the *Mathematica* **PACKAGES** directory. These two objects (**VISUAL** and **SETTINGS.MA**) will then be at the same level as the **ALGEBRA** directory or the **INIT.M** package, for example.

If you install **VISUAL** and **SETTINGS.MA** in a different place or choose to leave them on the CD-ROM, or if you don't have write access to the *Mathematica* directory, you must tell *Mathematica* where to find the packages before trying to read them.

If you want to use the packages directly from the CD-ROM (loaded in drive **D:**, for example), you can tell *Mathematica* where to find them by this command:

    $Path = Join[$Path, {"D:\\"}];

If you copy **VISUAL** and **SETTINGS.MA** into the subdirectory **PICTS** of **MATHE** on disk **C:**, for example, you should use

    $Path = Join[$Path, {"C:\\MATHE\\PICTS\\"}];

To have *Mathematica* perform this command automatically at startup, you can use any text editor to write this command into the **INIT.M** file in the *Mathematica* **PACKAGES** directory.

NeXTSTEP

There is no notebook reader (*MathReader*) for NeXT. Therefore, you need *Mathematica* to look at the notebooks.

Please follow installation instructions for *Mathematica* or ask your dealer for help to install *Mathematica*.

All *Illustrated Mathematics* files are contained in the compressed tar archive **NEXT_TAR.Z**.

Extracting the Tar Archive

Copy **NEXT_TAR.Z** into your home directory. Rename the copy **IlluMath.tar.Z**. Double-clicking it will cause it to be expanded into **IlluMath.tar**. Double-clicking this file will extract its contents into a **IlluMath** directory containing the complete *Illustrated Mathematics*. You may delete **IlluMath.tar** afterward.

Configuring *Mathematica* to Create Your Own Examples

Mathematica must find the **Visual** directory and the **Settings.ma** notebook. To allow the *Mathematica* kernel to find these programs, you must either install the programs in a place searched by *Mathematica*, or you must tell the kernel where to find them.

Copy into the Packages Directory

You can copy **Settings.ma** and **Visual** either into *Mathematica*'s **Packages** directory (**/LocalApps/Mathematica.app/Library/Mathematica/Packages**, if you installed *Mathematica* in **/LocalApps**), or into your own **Packages** directory, located in **~/Library/Mathematica/Packages**. (**~/Library/Mathematica/Packages** is being created the first time you start *Mathematica*.)

Append to *Mathematica's* Search Path

You can simply append the path of the **IlluMath** directory (where **Settings.ma** and **Visual** are located) to the variable **$Path** of *Mathematica*.

    AppendTo[$Path, "~/IlluMath"];

To have *Mathematica* perform this command automatically at startup, you should write this command into your **init.m** file. If you don't already have a **init.m** file in your home directory or in **~/Library/Mathematica/Packages**, you should copy the **init.m** file from the *Mathematica* **StartUp** directory into **~/Library/Mathematica/Packages**.

Use any text editor to add this line to **init.m**.

    AppendTo[$Path, "~/IlluMath"];

Use the correct path if you haven't installed *Illustrated Mathematics* in **~/IlluMath**. This directory must contain the **Settings.ma** file and the **Visual** directory.

You can check the correct path in **init.m** as follows:

If *Mathematica* is running, quit and restart it.

Open one of the notebooks in the **Collection** or **Manual** directory. Evaluate its initialization. It should evaluate without errors.

If there are errors, check the value of **$Path**.

$Path

{., ~, ~/Library/Mathematica/Packages,

/LocalApps/Mathematica22.app/Library/Mathematica/Packages,

/LocalLibrary/Mathematica/Packages,

/LocalApps/Mathematica22.app/Install/Preload,

/LocalApps/Mathematica22.app/StartUp, ~/IlluMath}

Installing in a Shell

You may also install *Illustrated Mathematics* from a shell. Read the section on Unix to learn how to do this.

Unix

For some Unix systems, a version of *MathReader* is now available, for example, for SunOS 4.1 and for Solaris 2. You can download a copy of *MathReader* from *MathSource*, either by anonymous ftp from **mathsource.wri.com** or through a WWW browser, such as Mosaic or Netscape at **http://www.wri.com/**.

If a separate notebook reader (*MathReader*) is not available for your system, you need a version of *Mathematica* that includes the notebook front end. For most computers supporting the X11 window system, such a front end is now available.

Please follow installation instructions for *Mathematica* or ask your dealer for help to install *Mathematica*.

All files of *Illustrated Mathematics* are contained in the compressed tar archive **UNIX_TAR.Z**. You may also use the notebooks of the Windows version instead. However, we recommend you unpack the Unix archive.

Mounting the CD-ROM

We assume that the CD-ROM should be mounted onto the directory **/cdrom**. If you use a different name, you should change all examples below accordingly. On some systems, such as Solaris 2, the CD-ROM will be mounted automatically after being inserted into the built-in CD-ROM drive.

Usually, only the superuser (root) can mount CD-ROMs. The command to do so varies from manufacturer to manufacturer. Under SunOS 4, the CD-ROM is mounted with this command.

/etc/mount -r -t hsfs /dev/sr0 /cdrom

Extracting the Tar Archive

Open a shell window (**xterm** or **cmdtool**). Go into your home directory or into the directory where you want to put the **IlluMath** directory.

Extract the contents of the archive **UNIX_TAR.Z**.

    `uncompress < /cdrom/unix_tar.z | tar xvf -`

Depending on the device driver, filenames will appear in either lowercase or uppercase letters.

An **IlluMath** directory with the complete *Illustrated Mathematics* will be created.

Partial Extractions

You may also extract parts of *Illustrated Mathematics* only, for example, the manuals.

    `uncompress < /cdrom/unix_tar.z | tar xvf - IlluMath/Manuals`

Configuring *Mathematica* to Create Your Own Examples

Mathematica must find the **Visual** directory and the **Settings.ma** notebook. To allow the *Mathematica* kernel to find the programs, you must either install the programs in a place searched by *Mathematica*, or you must tell the kernel where to find them.

Copy into the Packages Directory

You can copy **Settings.ma** and **Visual** either into *Mathematica*'s **Packages** directory, or into your home directory.

Append to *Mathematica*'s Search Path

You can simply append the path of the **IlluMath** directory (where **Settings.ma** and **Visual** are located) to the variable **$Path** of *Mathematica*.

    `AppendTo[ $Path, "~/IlluMath"];`

To have *Mathematica* perform this command automatically at startup, you should write this command into the **init.m** file in your home directory.

Modifying init.m

If you don't already have an **init.m** file in your home directory, you should copy the **init.m** file from the *Mathematica* **StartUp** directory into your home directory. Use any text editor to add this line to **init.m**.

    `AppendTo[ $Path, "~/IlluMath" ];`

Use the correct path if you haven't installed *Illustrated Mathematics* in **~/IlluMath**. This directory must contain the **Settings.ma** file and the **Visual** directory.

You can check the correct path in **init.m** as follows:

If *Mathematica* is running, quit and restart it.

Open one of the notebooks in the **Collection** or **Manual** directory. Evaluate its initialization. It should evaluate without errors.

If there are errors, check the value of **$Path**.

> **$Path**
>
> {., ~, /usr/local/math/Install/Preload,
>
> /usr/local/math/StartUp, /usr/local/math/Packages,
>
> ~/IlluMath}

Rendering the Notebooks

For reasons of disk space, the **.mb** files in the **Collection** and **Manuals** directories have been left out. If you open a notebook for the first time, the graphics will be rendered. This will take some time. When you save the notebook, the missing **.mb** file is created.

6 Notebook Documents

All documents on the *Illustrated Mathematics* CD-ROM are notebooks. Notebooks are hierarchically structured texts; they are also an interface to *Mathematica*.

This section describes how to look at notebooks. Notebooks can be displayed either with *MathReader* (enclosed on the CD-ROM), or with *Mathematica*. The explanations refer to *MathReader*. However, a note is given if the corresponding feature is different in *Mathematica* itself.

Mathematica allows you to create or edit notebooks, scale graphics, and so on. To learn more about this, refer to the documentation about *Mathematica*.

Groups: The Structure of Notebooks

A notebook is divided into chapters, section, subsections, etc., just like an ordinary book. Such a unit is called a group. The division into groups is made through the use of cells. There are title cells, text cells, graphics cells, and so on. A chapter group, for example, consists of a chapter cell (containing the title of the chapter) and all other cells in the chapter, including section cells, text cells, etc. The cells and groups are indicated by cell brackets at the right margin on the notebook window.

A group can be closed. Only the first cell in a closed group is visible (for example, the title of a chapter). All other cells are represented by a box.

A group is opened by double-clicking the box.

A group is closed by double-clicking the bracket that comprises the whole group.

Here is the notebook window on a Macintosh:

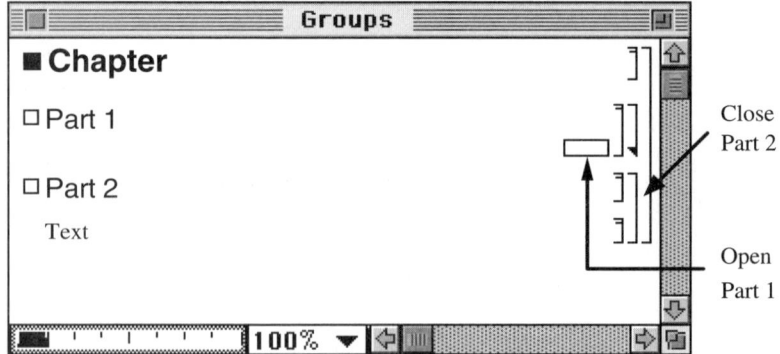

This is the notebook window under Windows:

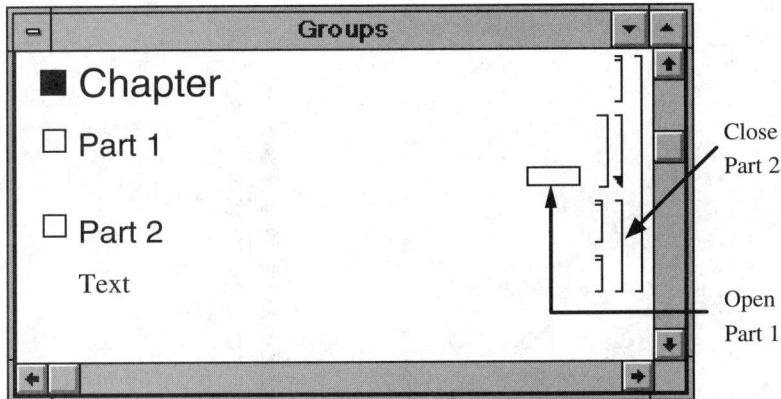

In larger documents all groups will usually be initially closed, except the outermost group containing all the chapters.

Graphics and Animations

Graphics are contained in graphics cells.

An animation is a sequence of graphics cells that are displayed one after the other, as in an animated film.

An animation is started by selecting the corresponding graphics (click on the enclosing cell bracket) and then choosing the **Animate Selected Graphics** item in the **Graph** menu.

Macintosh, NeXT, and X front end: An animation can also be started by double-clicking the first graphic.

After starting the animation, speed and direction can be changed by the palette of buttons displayed at the lower margin of the window. The first three buttons specify the direction of the animation: backward, back and forth, forward. The middle button is the pause button and the last two buttons adjust the speed: slower and faster.

This is an animation on the Macintosh:

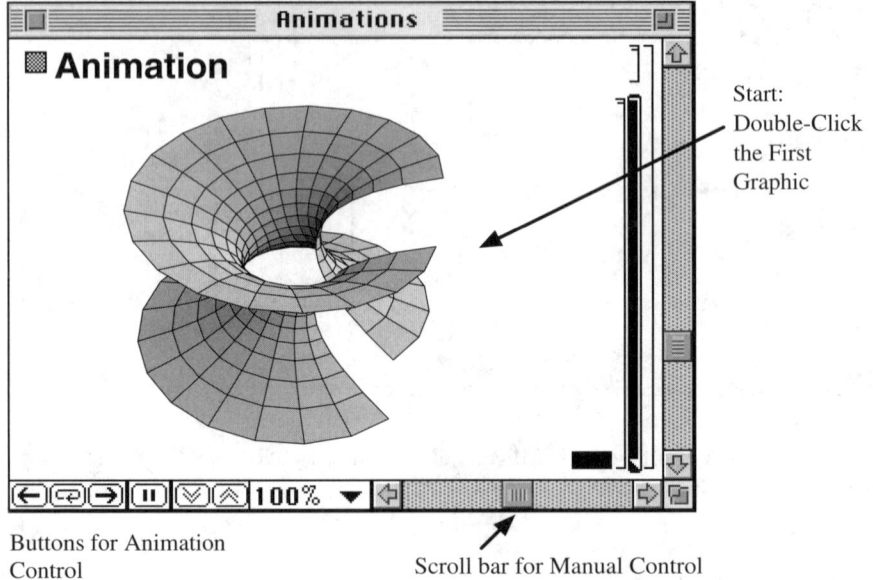

Buttons for Animation Control

Scroll bar for Manual Control

Start: Double-Click the First Graphic

Macintosh and NeXT: During an animation the horizontal scroll bar can be used to control the animation by hand.

An animation is stopped by any mouse click in the window.

Window Size and Display Size

Macintosh: The size of the notebook window is adjusted as usual at the lower-right corner. A click in the upper-right corner puts the window into a standard size, depending on the size of the monitor used.

Windows: The size of the notebook window is modified as usual. The size of individual graphics is adjusted by clicking the graphics and dragging the corners. To adjust the size of an animation, first adjust the size of the first graphic; then, use the **Align Selected Graphics** command in the **Graph** menu to set all graphics to the same size.

Macintosh and NeXT: The magnification (for texts and graphics) can be adjusted with the magnification box in the lower margin. Click the pop-up menu with the percentage display to choose from among several possibilities.

If the magnification of the notebook or the color palette of your computer is modified, graphics are not redrawn automatically (this could occupy a computer for several minutes). To redraw the graphics, click the desired graphics and choose the **Render PostScript** or **Re-render PostScript** item in the **Graph** menu. To redraw all graphics in a group (for example, all graphics in the notebook), click the desired group and then issue the command as above. This feature is not available under Windows.

To allow the use of the graphics on computers with a small amount of random access memory (RAM), several notebooks have been stored with converted graphics. These converted graphics are stored using certain bitmap formats that cannot scale easily.

Printing

The **File** menu contains the **Print** command. It can be used to print a whole notebook or single pages. Before printing, you should choose the correct page size (for example, US letter) in the **Setup** menu.

Macintosh: **Page Setup** is in the **File** menu (in the **Printing Settings** menu in *Mathematica*).

Windows: **Setup** can be selected in the **Print** menu.

The *Illustrated Mathematics* notebooks have been formatted for vertically oriented letter-sized paper.

If you want to print only a part of a notebook, you can close the remaining groups.

Macintosh and NeXT: If you want to eliminate individual cells (including first cells of closed groups) you can do this as follows. The **Cell** menu contains the **Closed** item (among **Attributes** in the **Style** menu in *Mathematica*). This command closes the cells and reduces them to a small bracket in the right margin. Selecting **Closed** again opens the cells. (In *Mathematica* the command **Print Selection** in the **File** menu prints selected cells only).

Windows: Selecting **Selection** in the **Print** menu causes only selected cells to be printed.

To allow the use of the graphics on computers with a small amount of RAM, several notebooks have been stored with converted graphics. These converted graphics are stored using certain bitmap formats that can be printed at screen resolution only.

Copying Graphics into Other Documents

MathReader doesn't enable you to copy graphics out of notebooks. To allow you to use the graphics in other programs, many of our graphics have been stored separately in the **Images** directory/folder. You can open these single images with many programs.

Macintosh: If your word processor cannot open the images directly, you can open them with *TeachText/SimpleText* (these programs are part of the operating system) and put the images or parts of them onto the Clipboard.

Windows: The images were saved in the *Windows Meta-File* format (extension ***.wmf**). Many programs, such as *Word* or *Corel*, support this format.

Further information can be found in the **ReadMe.ma** file in the **Images** directory.

Mathematica allows you to save graphics in many formats or put them onto the Clipboard. You can also create your own notebooks.

Run-Time Problems

Currently, CD-ROM drives are much slower than hard disks. You may want to copy frequently used notebooks onto a hard disk.

Macintosh: Note that many notebooks in the collection have been stored with converted graphics as well. This format allows you to open them with a smaller amount of memory (RAM).

The first iteration through an animation is often slower, since all data has to be read first.

Windows: To save space, some notebooks have been stored without the binary information. If such a notebook is opened, the missing data are restored automatically, including bitmaps for graphics displayed on the screen. When giving a demonstration, you may want to open the notebook beforehand to allow the rendering of graphics to complete before you use the notebook. If you use *Mathematica*, the binary information will be saved when you save the notebook.

If you have trouble opening a notebook (e.g., if you use another version of *Mathematica*), you may try to open it with the **No Binary** box checked.

Macintosh: If you have a large amount of RAM or use virtual memory, you can improve the performance of *MathReader* by giving it more memory. To do so you must copy *MathReader* onto a hard disk. See the Installation section of the booklet or the **Installation.ma** notebook for more details.

If you run out of memory looking at a notebook, you may try to close and reopen the notebook.

7 Using the Programs

This section gives a short introduction to the use of the programs on the CD-ROM. Please note; that you need *Mathematica* and that the programs must be installed correctly, see the Installation section of the booklet or the **Installation.ma** notebook for more details.

The simplest way to use the programs is to modify an example in an existing notebook. To do so, open a notebook with examples for the command you want to use. Usually, you will be asked whether you want to evaluate the initialization of the notebook you opened. After that, you can modify parameters in existing examples and generate new ones.

A complete description of the commands of a program package can be found in the corresponding manual.

Mathematica's User Interface

There are cells for *Mathematica* input and output. To perform a computation, select the input cell (click on the bracket or anywhere inside the cell); then, press **Shift-Return** (for the Macintosh **Enter** works as well). Output is added right after the input cell (old output is removed first).

Loading Programs: Needs

The programs must be loaded *before* they can be used. Programs are loaded with the command **Needs**.

    Needs["Visual`Package`"];

According to the installation instructions, the programs can be found in the **Visual** directory. Insert the correct name for **Package** (without the extension .m).

If you try to use a command before its package has been loaded, the command will not be executed. In this case you must quit *Mathematica*, restart it, and load the package before trying to evaluate the command.

If you begin with one of the *Illustrated Mathematics* notebooks, the cells with the commands to load the packages are marked as initialization cells. This means that you will usually be asked whether you want to evaluate these commands when you open the notebook.

Documentation

All the commands are documented as usual. With

   ?Command

you obtain the usage message for *Command*. The corresponding package must have been loaded beforehand.

Macintosh and NeXT: The **Function Browser** (in the **Help** menu) gives you usage messages and templates for using the commands. Such templates can be copied into the notebook.

8 General Remarks about *Mathematica*

This section explains some principles of *Mathematica*.

Syntax

Mathematica evalutes input expression according to well-defined rules. All our programs are nothing more than rules for the evaluation of certain expressions (commands).

A simple expression is a number, a string, or a symbol.

A normal expression has the form

    h [a, b, ...]

where **h** is the head of the expression, and **a, b,** ... are the arguments (or elements). White space can be used to improve readability. Note that square brackets are used, not ordinary parentheses.

    Sin[x]

The multiplication operator ***** can be left out. Round parentheses are used only for grouping expression (as they are in mathematics).

    a (b + c)

Lists are written with curly braces.

    {x, a, b}

Lists are used also to indicate a range.

    Plot[x^2, {x,-2,2}];

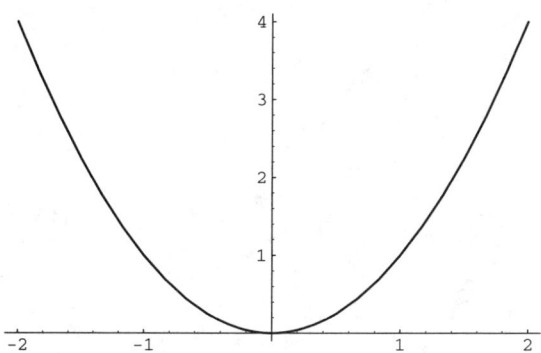

In the preceding example, the list is read as "... for **x** in the interval [-2, 2] ... ". A semicolon **;** at the end of input supresses textual output. This is often desired for graphics, since we are interested only in the graphics and not in its internal data structure (abbreviated by -Graphics-).

The names of all predefined functions and commands in *Mathematica* have capital initial letters.

Mathematica evaluates expressions until no more rules can be applied. The result is then output.

```
(3 + 5)/4 + Sin[Pi] + Cos[Pi/4+Pi/9]
          13 Pi
2 + Cos[ ───── ]
           36
```

A percent sign `%` represents the last result, `%%` represents the next-to-last result, and so on.

```
Plot[ Sin[x], {x, 0, 2 Pi}];
```

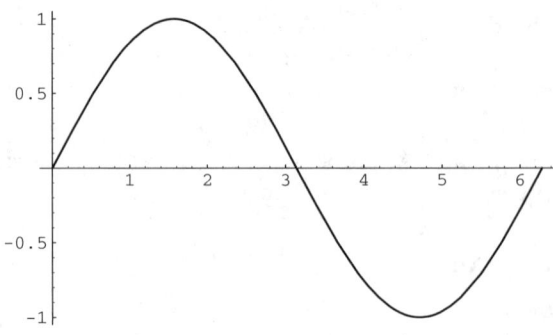

```
Plot[ Cos[x], {x, -Pi/2, 3 Pi/2}];
```

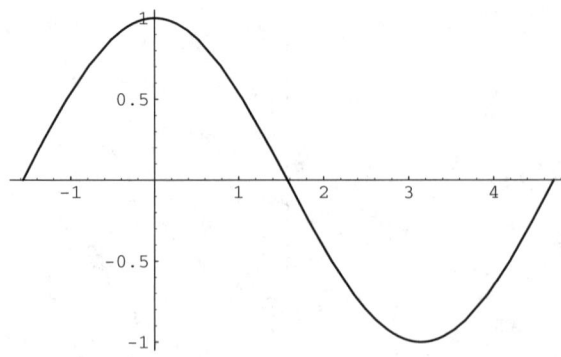

```
Show[ %, %%];
```

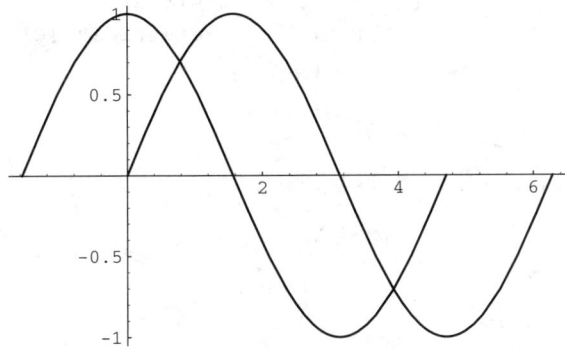

Options

Many commands (including those in *Illustrated Mathematics*) can take so-called options, having the form

    `Parameter -> Value`

Options are used to set additional parameters that are not available as ordinary arguments of a command. Options can be added in arbitrary order at the end of an argument list.

    `Command [ arg1, arg2, ..., argn, par -> val, ... ]`

This mechanism offers great flexibility, especially for graphics commands. It enables a simple specification of such details as axes labels, plot range, and so on.

```
ParametricPlot[ {Sin[2t] Cos[t], Sin[2t] Sin[t]},
                {t, 0, 2 Pi},
                PlotLabel -> "r = Sin(2t)",
                DefaultFont -> {"Helvetica",12},
                Axes -> None ];
```

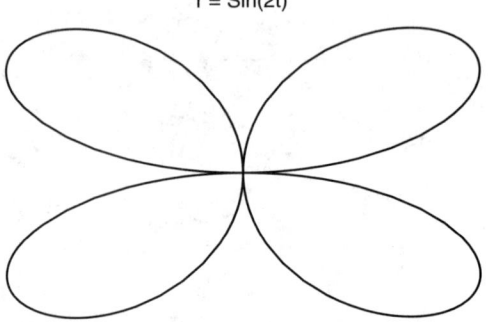

The command
> **Options[** *Command* **]**

gives a list of all options of *Command*.

9 Graphics with *Mathematica*

This section describes how to work with *Mathematica* graphics. It contains also a summary of the built-in graphics capabilities of *Mathematica*.

Graphic Commands

You find a complete description of the built-in graphic commands in the *Mathematica* book (Introduction 1.9, Principles 2.9, Reference Guide A.8).

The following *Mathematica* commands can be used to create visualizations of data.

    ListPlot, ListPlot3D,
    ListContourPlot, ListDensityPlot

To plot functions you can use the following commands.

    Plot, Plot3D,
    ContourPlot, DensityPlot,
    ParametricPlot, ParametricPlot3D

Plot

    ?Plot

Plot[f, {x, xmin, xmax}] generates a plot of f as a function
   of x from xmin to xmax. Plot[{f1, f2, ...}, {x, xmin,
   xmax}] plots several functions fi.

Plot is used to plot the graph of a unary function.

    Plot[Sin[x], {x, 0, 2 Pi}];

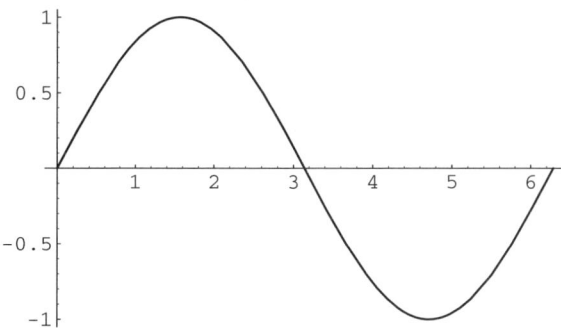

Plot3D

?Plot3D

Plot3D[f, {x, xmin, xmax}, {y, ymin, ymax}] generates a
   three-dimensional plot of f as a function of x and y.
   Plot3D[{f, s}, {x, xmin, xmax}, {y, ymin, ymax}] generates
   a three-dimensional plot in which the height of the
   surface is specified by f, and the shading is specified by s.

Graphs of binary functions *f*, which have two independent variables *x* and *y*, can be plotted by **Plot3D** as a surface in space ($z = f(x,y)$).

Plot3D[Sin[Sqrt[x^2+y^2]], {x, 0, 2 Pi}, {y, 0, 2 Pi}];

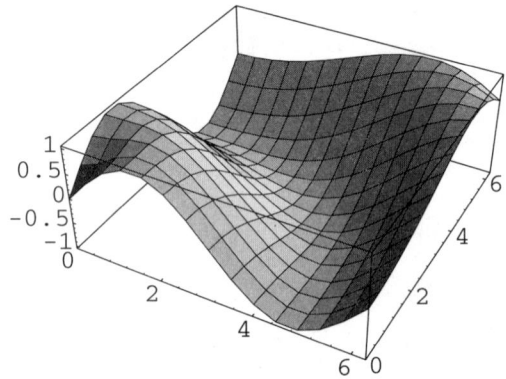

ContourPlot

?ContourPlot

ContourPlot[f, {x, xmin, xmax}, {y, ymin, ymax}] generates a
   contour plot of f as a function of x and y.

Another possibility to visualize graphs of binary functions is **ContourPlot**. Contour lines connect points with the same function value. The range in between is colored according to the function value: the lighter the color, the larger the value, and vice versa.

```
ContourPlot[ Sin[ Sqrt[x^2+y^2]],
            {x, 0, 2 Pi}, {y, 0, 2 Pi}];
```

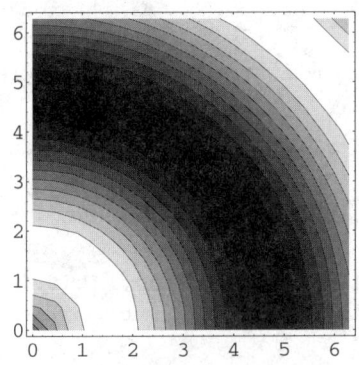

DensityPlot

```
?DensityPlot
```

DensityPlot[f, {x, xmin, xmax}, {y, ymin, ymax}] makes a
    density plot of f as a function of x and y.

Instead of plotting contour lines we can simply color a lattice: the darker the color of a field, the smaller the value of the function, and vice versa.

```
DensityPlot[ Sin[ Sqrt[x^2+y^2]],
            {x, 0, 2 Pi}, {y, 0, 2 Pi}];
```

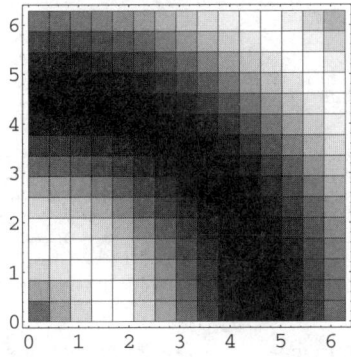

ParametricPlot

?ParametricPlot

```
ParametricPlot[{fx, fy}, {t, tmin, tmax}] produces a
   parametric plot with x and y coordinates fx and fy
   generated as a function of t. ParametricPlot[{{fx, fy},
   {gx, gy}, ...}, {t, tmin, tmax}] plots several parametric
   curves.
```

Curves of the form $y = f(x)$ can be drawn by **Plot**. You can use **ParametricPlot** to plot curves given in parametric representation $(x(t), y(t))$.

For example, this is a circle (the option causes the coordinate axes having the same scales):

```
ParametricPlot[ {Cos[t], Sin[t]}, {t,0,2 Pi},
                AspectRatio -> Automatic ];
```

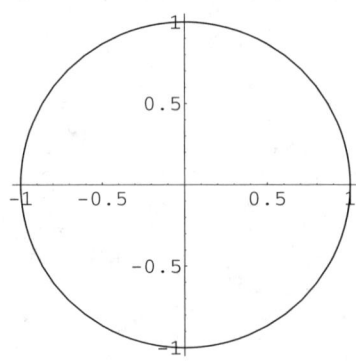

ParametricPlot3D

?ParametricPlot3D

```
ParametricPlot3D[{fx, fy, fz}, {t, tmin, tmax}] produces a
   three-dimensional space curve parameterized by a variable
   t which runs from tmin to tmax. ParametricPlot3D[{fx, fy,
   fz}, {t, tmin, tmax}, {u, umin, umax}] produces a
   three-dimensional surface parametrized by t and u.
   ParametricPlot3D[{fx, fy, fz, s}, ...] shades the plot
   according to the color specification s.
   ParametricPlot3D[{{fx, fy, fz}, {gx, gy, gz}, ...}, ...]
   plots several objects together.
```

For surfaces as well as for space curves given in parametric representation you can use **ParametricPlot3D** to create three-dimensional graphics.

This is a spiral in space:
ParametricPlot3D[{Cos[t], Sin[t], t/10}, {t, 0, 6 Pi}];

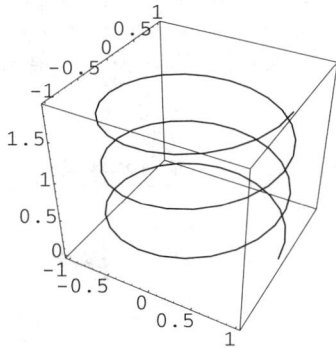

This is a sphere:
**ParametricPlot3D[{Cos[t] Cos[p], Cos[t] Sin[p], Sin[t]},
               {p, -Pi, Pi}, {t, -Pi/2, Pi/2}];**

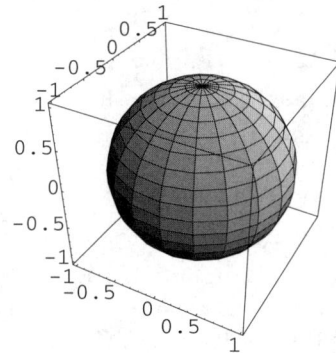

Graphic Objects

Each of the graphic commands mentioned above not only creates a graphic on the screen but creates an internal object of the graphic as well. The internal representation allows us to change parameters afterward or to put several graphics together.

Types of Graphics

Each graphic has a type, which depends on the graphic command used to create it. (If you drop the semicolon ; at the end of the input, the type of the graphic is printed in an additional output line below the graphic.)

These are the graphic commands and their corresponding types:

>   Plot
>   ListPlot
>   ParametricPlot   Graphics
>
>   Plot3D
>   ListPlot3D   SurfaceGraphics
>
>   ContourPlot
>   ListContourPlot   ContourGraphics
>
>   ListDensityPlot
>   DensityPlot   DensityGraphics
>
>   ParametricPlot3D   Graphics3D

Composition of Graphics

We use **Show** to put several graphics together into one:

>   Show[g1, g2, ..., gn]

Options

Parameters, which are not ordinary arguments and which determine details such as labels, lighting, colors, viewpoint, frame, and so on, are set by options. Most of these options can be used in the *Illustrated Mathematics* commands as well.

Options are always of the form:

>   *Parameter* -> *Value*

They can be added in arbitrary order at the end of the normal argument list:

>   *Command* [*arg1*, *arg2*, ..., *argn*, *par* -> *val*, ...]

With

>   Options[*Command*]

you obtain a list of the possible options of a *Command*.

Options[DensityPlot]

{AspectRatio -> 1, Axes -> False, AxesLabel -> None,

　AxesStyle -> Automatic, Background -> Automatic,

　ColorFunction -> Automatic, ColorOutput -> Automatic,

　Compiled -> True, DefaultColor -> Automatic,

　Epilog -> {}, Frame -> True, FrameLabel -> None,

　FrameStyle -> Automatic, FrameTicks -> Automatic,

　Mesh -> True, MeshStyle -> Automatic, PlotLabel -> None,

　PlotPoints -> 15, PlotRange -> Automatic,

　PlotRegion -> Automatic, Prolog -> {},

　RotateLabel -> True, Ticks -> Automatic,

　DefaultFont :> $DefaultFont,

　DisplayFunction :> $DisplayFunction}

In this list each parameter is set to the default value that is used by *Mathematica*. These values are either given explicitly or by one of the following forms:
Automatic: an internal algorithm is used to set the parameter.
　None: the parameter is not handled.
　All: all points, etc. are to be included.
　True: the parameter is enabled.
　False: the parameter is disabled.

Many parameters affect the output of the graphic but not the computation of its internal object. We use **Show** to change these parameters afterward (without computing the internal object again). At the same time, we can compose different graphics with **Show**.

　?Show

```
Show[graphics, options] displays two- and three-dimensional
    graphics, using the options specified. Show[g1, g2, ...]
    shows several plots combined. Show can also be used to
    play Sound objects.
```

graph1 = Plot[Sin[x], {x, 0, 2 Pi}];

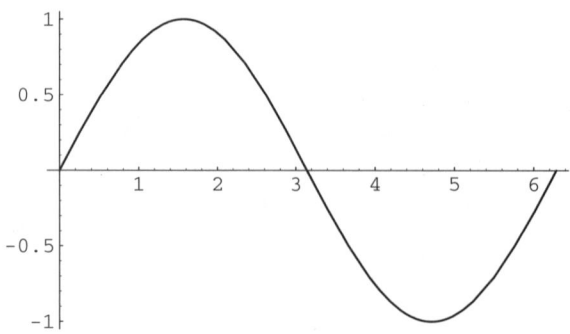

Show[graph1, Axes -> None, Frame -> True];

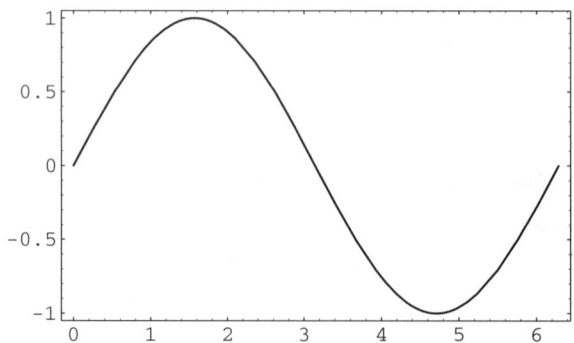

Every type of graphic has its own options. These options can be modified at the internal object. The command

Options[*Type*]

prints the list of options for the graphics of type *Type*.

```
Options[Graphics]
                          1
{AspectRatio -> ───────────, Axes -> False,
                      GoldenRatio

  AxesLabel -> None, AxesOrigin -> Automatic,

  AxesStyle -> Automatic, Background -> Automatic,

  ColorOutput -> Automatic, DefaultColor -> Automatic,

  Epilog -> {}, Frame -> False, FrameLabel -> None,

  FrameStyle -> Automatic, FrameTicks -> Automatic,

  GridLines -> None, PlotLabel -> None,

  PlotRange -> Automatic, PlotRegion -> Automatic,

  Prolog -> {}, RotateLabel -> True, Ticks -> Automatic,

  DefaultFont :> $DefaultFont,

  DisplayFunction :> $DisplayFunction}
```

PlotPoints, PlotDivision

```
?PlotPoints
```
PlotPoints is an option for plotting functions. With a single
   variable, PlotPoints -> n uses a total of n sample points.
   With two variables, PlotPoints -> n specifies that n
   points should be used in both the x and y directions, and
   PlotPoints -> {nx, ny} specifies different numbers of
   sample points for the two directions.

```
?PlotDivision
```
PlotDivision is an option for Plot. PlotDivision -> n
   specifies that a maximum of n subdivisions should be used
   in attempting to generate a smooth curve.

PlotPoints is used to determine the number of sample points evaluated for curves and surfaces. In the case of a plane curve, **PlotDivision** is used to define the maximum number of additional subdivisions.

Both parameters affect time and memory use as well as the resolution of the graphic. Therefore, they cannot be changed with **Show** after the computation of the internal object.

PlotRange

> **?PlotRange**

PlotRange is an option for graphics functions. With PlotRange
  -> All, all points are included in the plot. With
  PlotRange -> Automatic, outlying points are dropped.
  PlotRange -> {min, max} specifies explicit limits for y
  (2D) or z (3D). PlotRange -> {{xmin, xmax}, ... } gives
  explicit limits for each coordinate.

> **Plot[1/x, {x,-4,4}, PlotRange -> {{-5,5},{-3,3}}];**

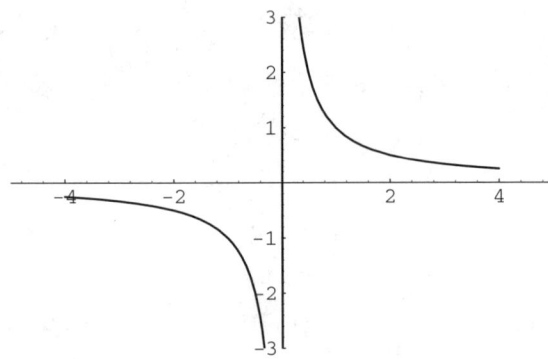

This parameter can also be changed afterward by **Show**.

AspectRatio

> **?AspectRatio**

AspectRatio is an option for Show and related functions. With
  AspectRatio -> Automatic, the ratio of height to width of
  the plot is determined from the actual coordinate values
  in the plot. AspectRatio -> r makes the ratio equal to r.

AspectRatio -> Automatic causes the scales of the coordinates to be equal.

This parameter can also be changed afterward by **Show**.

ViewPoint

?ViewPoint

ViewPoint is an option for Graphics3D and SurfaceGraphics which gives the point in space from which the objects plotted are to be viewed. ViewPoint -> {x, y, z} gives the position of the view point relative to the center of the three-dimensional box that contains the object being plotted.

Relative coordinates are used: **{0,0,0}** is the center point of the bounding box; the length of the box is one unit.

**Plot3D[Exp[-(x^2+y^2)], {x,-2,2},{y,-2,2},
        ViewPoint -> {2, -2.5, 0.4}];**

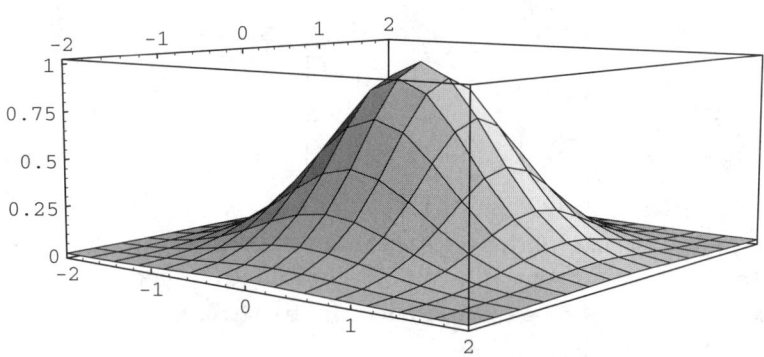

The notebook front ends (Windows, Macintosh, and NeXT) can help you find the desired viewpoint. Choose **3D ViewPoint Selector** from the **Prepare Input** submenu of the **Action** menu. Then, you can set your viewpoint by turning a box or moving scroll bars. You can directly paste the determined values into the actual input cell of the notebook.

This parameter can also be changed afterward by **Show**.

Axes, Frame, Box, and GridLines

Examples:
  Axes -> None: No axes are drawn.
  AxesOrigin -> {x,y}: The axes intersect at (*x,y*).
  Frame -> True: A frame is drawn around the graphic.
  Boxed -> False: The box around the three-dimensional graphic is omitted.

Further related parameters include: **Ticks**, **AxesEdge**, **BoxRatio**, **FrameTicks**, **GridLines**, and **FaceGrids**. These parameters can also be changed afterward by **Show**.

Labels

Example:
```
Plot[ Tan[x], {x, 0, 2 Pi},
      AxesLabel -> {"Arc",None},
      PlotLabel -> "The Tangent",
      DefaultFont -> {"Times",9} ];
```

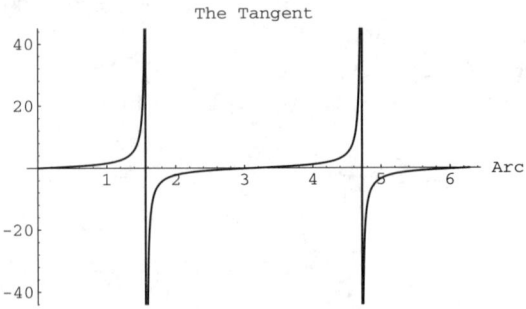

Parameters for labeling include: **PlotLabel**, **AxesLabel**, **FrameLabel**, **RotateLabel**. **DefaultFont** is used to determine the font for text.
These parameters can also be changed afterward by **Show**.

Parameter Lines of Three-dimensional Graphics

You can eliminate the parameter lines of a three-dimensional graphic (type **SurfaceGraphics** or **Graphics3D**) by **Mesh -> False**.
```
Plot3D[ Sin[x] Sin[y], {x,0,2 Pi}, {y,0,2 Pi},
        Mesh -> False ];
```

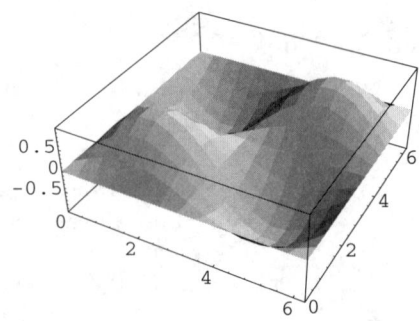

This option can be used for **DensityGraphics** as well.

This parameter can also be changed afterward by **Show**.

Lighting of Three-dimensional Graphics

The parameters **AmbientLight**, **Lighting**, and **LightSources** allow you to modify the lighting model used for three-dimensional graphics.

Important Parameters of Contour Graphics

?Contours

```
Contours is an option for ContourGraphics specifying the
    contours to use. Contours -> n chooses n equally spaced
    contours between the minimum and maximum z values.
    Contours -> {z1, z2, ... } specifies the explicit z values
    to use for contours.
```

?ContourShading

```
ContourShading is an option for contour plots. With
    ContourShading -> False, regions between contour lines are
    left blank. With ContourShading -> True, regions are
    colored based on the setting for the option ColorFunction.
```

?ContourLines

```
ContourLines is an option for contour plots. With
    ContourLines -> True, explicit contour lines are drawn.
    With ContourLines -> False, no contour lines are drawn.
```

10 Selected Publications about *Mathematica*

Introductory

Blachman, Nancy. *Mathematica: A Practical Approach*, Prentice Hall, 1992. ISBN 0-13-563826-7

Gray, Theodore, and Jerry Glynn. *The Beginner's Guide to Mathematica Version 2*, Addison-Wesley, 1992. ISBN 0-201-58221-X

Shaw, William T., and Jason Tigg. *Applied Mathematica: Getting Started, Getting it Done*, Addison-Wesley, 1994. ISBN 0-201-54217-X

Wagon, Stan. *Mathematica in Action*, W. H. Freeman, 1991. ISBN 0-7167-2229-1

Handbooks/Reference

Wolfram, Stephen. *Mathematica: A System for Doing Mathematics by Computer*, Second Edition, Addison-Wesley, 1991. ISBN 0-201-51502-4

Wolfram, Stephen. *The Mathematica Reference Guide*, Addison-Wesley, 1992.

Guide to Standard Mathematica Packages, Wolfram Research Technical Report, 1991. ISBN 1-880083-09-4

Wolfram, Stephen, with George Beck. *Mathematica: the Student Book*, Addison-Wesley, 1994.

Trott, Michael. *The Mathematica Guidebook*, TELOS/Springer-Verlag, 1995. Includes CD-ROM. ISBN 0-387-94282-3

Wickham-Jones, Tom. *Mathematica Graphics: Techniques and Applications*, TELOS/Springer-Verlag, 1994. ISBN 0-387-94047-2

Programming

Gaylord, Richard J.; Kamin, Samuel N.; and Paul R. Wellin. *Introduction to Programming with Mathematica*. TELOS/Springer-Verlag, 1993. ISBN 0-387-94048-0

Gray, John W. *Mastering Mathematica: Programming Methods and Applications*, AP Professional, 1994. ISBN 0-12-296040-8

Maeder, Roman E. *Programming in Mathematica*, Second Edition, Addison-Wesley, 1991. ISBN 0-201-54877-1

Maeder, Roman E. *The Mathematica Programmer*, AP Professional, 1994. ISBN 0-12-464990-4

Applications

Gaylord, Richard J., and Paul R. Wellin. *Computer Simulations with Mathematica*, TELOS/Springer-Verlag, 1994. With CD-ROM. ISBN 0-387-94274-2

Gray, Theodore, and Jerry Glynn. *Exploring Mathematics with Mathematica*, Addison-Wesley, 1991. ISBN 0-201-52809-6

Skeel, R., and Jerry Keiper. *Elementary Numerical Computing with Mathematica*, McGraw-Hill, 1993. ISBN 0-07-057820-6

Skiena, Steven S. *Implementing Discrete Mathematics: Combinatorics and Graph Theory with Mathematica*, Addison-Wesley, 1990. ISBN 0-201-50943-1

Varian, Hal R. (ed.). *Economic and Financial Modeling with Mathematica*, TELOS/Springer-Verlag, 1993. ISBN 0-387-97882-8

Vvedensky, Dimitri. *Partial Differential Equations with Mathematica*, Addison-Wesley, 1993. ISBN 0-201-54409-1

Teaching

Davis, Bill; Porta, Horacio; and Jerry Uhl. *Calculus&Mathematica*, Addison-Wesley, 1994. ISBN 0-201-58461-1 (Windows), 0-201-58153-1 (Macintosh/NeXT).

Crooke, Philip, and John Ratcliffe. *A Guidebook to Calculus with Mathematica*, Wadsworth Publishing Company, 1991.

Braden, Bart. *Discovering Calculus with Mathematica*, John Wiley&Sons, 1992.

Finch, James K., and Millanne Lehmann. *Exploring Calculus with Mathematica*, Addison-Wesley, 1992. ISBN 0-201-55572-7

Periodicals

The Mathematica Journal, Miller Freeman, Inc. ISSN 1047-5974

Mathematica in Education, Paul Wellin, editor, (1994), TELOS/Springer-Verlag. ISSN 1065-2965

Mathematica World (floppy disk magazine), Stephen M. Hunt, editor. (also available on CD-ROM)

11 Further Development of *Illustrated Mathematics*

We welcome comments, hints, and bug reports. Please direct your correspondence to the publisher or by email to the authors.